Nature Journal

Philip Henry Gosse as a young man of
twenty-nine, the year of his return to
England from Alabama, painted by his
brother, William Gosse. (1839, watercolor
on ivory, courtesy of the National Portrait
Gallery—London)

ADVISORY PANEL

Dr. Gary R. Mullen, Entomology, Auburn University

Dr. L. J. Davenport, Botany, Samford University

Elberta Gibbs Reid, Birmingham Audubon Society

Dr. E. O. Wilson, Zoology, Harvard University (Emeritus)

INAUGURAL SPONSORS

Birmingham Audubon Society

Mary Carolyn Gibbs Boothby

Lida Inge Hill

Fay Belt Ireland

Henry S. Lynn Jr.

Medical Management Plus, Inc.

Dr. Connie S. and Dr. James A. Pittman Jr.

Elberta Gibbs Reid

Frances B. and James W. Shepherd

Becky H. and William E. Smith Jr.

Virginia B. and William M. Spencer III

Ann A. and Donald B. Sweeney Jr.

Dr. Cameron M. and Judge J. Scott Vowell

Alice M. and N. Thomas Williams

Harriett Harton Wright

Louise A. and John N. Wrinkle

Philip Henry Gosse (1810–1888) was an English naturalist and illustrator who spent eight months of 1838 on the Alabama frontier, teaching planters' children in Dallas County and studying the native flora and fauna. Years after returning to England, he published the now-classic *Letters from Alabama, (U.S.) Chiefly Relating to Natural History* (1859), with twenty-nine important black-and-white illustrations included. He also produced, during his Alabama sojourn, forty-nine remarkable watercolor plates of various plant and animal species, mainly insects.

The Gosse Nature Guides are a series of natural history guidebooks prepared by experts on the plants and animals of Alabama and designed for the outdoor enthusiast and ecology layman. Because Alabama is one of the nation's most biodiverse states, its residents and visitors require accurate, accessible field guides to interpret the wealth of life that thrives within the state's borders. The Gosse Nature Guides are named to honor Philip Henry Gosse's early appreciation of Alabama's natural wealth and to highlight the valuable legacy of his recorded observations. Look for other volumes in the Gosse Nature Guides series at http://uapress.ua.edu.

The University of Alabama Press   Tuscaloosa

L. J. Davenport

With a Foreword by John C. Hall

Nature Journal

11 July 2011
to Ryan, with
all good wishes

The University of Alabama Press
Tuscaloosa, Alabama 35487-0380
All rights reserved
Printed in China

Typeface: Garamond Premiere Pro
Designed by Michele Myatt Quinn

∞
The paper on which this book is printed meets the minimum
requirements of American National Standard for Information
Sciences-Permanence of Paper for Printed Library Materials,
ANSI Z39.48-1984.

Library of Congress Cataloging-in-Publication Data

Davenport, Lawrence James.
　　Nature journal / L. J. Davenport; with a foreword by John C. Hall.
　　　　p.　cm. — (Gosse nature guides)
　　ISBN 978-0-8173-5569-2 (pbk. : alk. paper) — ISBN 978-0-8173-
8391-6 (electronic) 1. Natural history — Alabama. I. Title.
　　QH105.A2D38 2010
　　508.761—dc22　　　　　　　　　　　　　2009035170

Publication supported in part by Samford University

Cover: Great Blue Heron (*Ardea Herodias*), original waterolor painting
(34 x 25¼ inches) by Basil Ede, 1978. Property of the Westervelt
Corporation, Tuscaloosa, AL.

For W. Mike Howell—mentor, colleague, collaborator, and friend

"'Careful and minute descriptions, accurate measurements, and distinctive names are absolutely indispensable to science . . . but they must not be confounded with science itself.' . . . A true knowledge of the natural world comes only from getting out into it, from climbing trees and mountains, wading along creeks, along the shore and out into the sea, from crouching hidden in tangled creepers. It comes from 'protracted acquaintance' with the animals, the insects, the reptiles, the bats and birds in the place where they live."

—Philip Henry Gosse, as quoted and paraphrased by Ann Thwaite in *Glimpses of the Wonderful: The Life of Philip Henry Gosse*

Contents

Foreword

It is comforting to see the growing body of accessible books concerning the natural history of Alabama and the mid-South. The reported wonders of the Alabama environment are not just propaganda from various environmental organizations. The state really is among the most interesting and diverse environments in North America, a world-class landscape made ordinary only by its familiarity to us Alabamians. Its many international rankings are in number of fishes, variety of reptiles and amphibians, diversity of snails and mussels, number of streams, variety of geology and landscapes, diversity of fossils (with the most dinosaurs in eastern North America), and in the sheer number of plants.

In this book, L. J. Davenport takes us on a wide-ranging tour of that natural history. His broad sense of science extends to the physical landscape and to the human story that runs through it. Those lucky enough to have read his work in *Alabama Heritage* will recognize his style, which creates a book that everyone will enjoy. Davenport makes his points clearly and succinctly without the cloud of jargon that so commonly accompanies science writing. And his broad view of natural history is a comfort to non-specialists—science made understandable without compromising its exacting nature. I'm torn to say which amazed me more, his account of the utter transformation of giant swallowtails and luna moths or the mathematical subdivisions of the periodical cicadas.

But this is more than a collection of clever essays. It is a series of challenges for naturalists, young and old, not to be just spectators of natural history, but to become full participants. Davenport challenges us to become members of the scientific team—observing, recording, and questioning—

because it's critical that understanding and appreciating these wonders not be just the realm of specialists. Amateurs, too, have important work to do in exploring and protecting this amazing place.

Why do you suppose they asked us to draw all that stuff in biology lab? There were plenty of diagrams in the book and, goodness knows, we weren't very good at drawing. But those of us who tried, particularly those who wound up teaching, fully understand. The point was never to make a good drawing of a crawdad; it was to force us to look hard enough to catch that unusual bend of the tail, the long spike between the eyes, those bumps on the claws, and all those little modified legs. Experienced teachers know you cannot truly understand something until you try to record and explain it. Davenport demands that we share with him the higher end of the process, not just the stamp-collecting of facts and species, but the telling.

Yes, it is hard to put our thoughts down on paper, to describe what we feel and think: the joy in seeing an enormous pileated woodpecker up close, the unexpected strike of a heron and its brief adjustment of the fish for the slippery headfirst trip down its gullet, the lizard doing his blue-bellied push-ups for love and territory. But the reward is to gain new insight into the problem, to learn details and connections.

It has been observed that to know something of value is to begin to love it. When enough of us love it, we will take care of it. It is this attitude that is going to make natural history a topic of thought and discussion for ordinary people, the folks who are going to demand that caring for the environment is not optional but a fundamental feature of citizenship.

—JOHN C. HALL

Acknowledgments

This book began twenty years ago when I was asked by Guy Hubbs, then an assistant editor of *Alabama Heritage,* to turn my previous (dry but scholarly) work on the botanist Charles Mohr into a popular magazine article. I struggled. But finally, and with the help and patience of both Guy and the magazine's founding editor, Suzanne Wolfe, I produced something we could all live with—factually accurate and at least moderately entertaining.

Guy and I next teamed up on a pictorial essay based on my early work with the Cahaba lily. And then Suzanne asked me to write a quarterly nature column. She saw a creative streak in me that I didn't know I had, then gave me a forum to express it. For that I will always be grateful.

With Suzanne's retirement, her work was ably picked up by Donna Cox Baker. And, over the years, a slew of assistant editors and interns encouraged me to take chances, prodded me to meet my deadlines, and picked at my final products. I learned much from all of you.

After the individual pieces had been written, Beth Motherwell at The University of Alabama Press stepped in to give this book its final form and emphasis—a "working" journal rather than a mere collection. I thank her for her vision and for her unwavering energy in steering this book to its completion.

Finally, I thank Samford University for two essential items: generous financial support in the final publication of this book and financial, moral, and other support during the fifteen years of its writing. The deans of the Howard College of Arts and Sciences, Rod Davis and David Chapman, and Provost Brad Creed are especially singled out for my thanks.

Nature Journal

Introduction

I confess to having no formal training in nature writing. Instead, my training is in field biology—observing natural phenomena, collecting information, and connecting to a greater whole. The writing always comes second, based entirely on the first.

Such writing is an extension of my teaching. As in any type of teaching, I collect all pertinent information, digest it down to a meaningful outline, then present it as a story, from beginning to middle to end. Depending on the topic, the final form may be serious, painfully personal, or light-hearted and even goofy. But the main point is to tell the story, and to tell it accurately and well.

This book is based on my "Nature Journal" columns, each of which appeared in *Alabama Heritage* between 1993 and 2008. But it's not just a collection of essays. Instead, each chapter is designed to stimulate you to make your own observations, collect your own information, and present your own findings.

This is an active book. The essay in each chapter is just one example of how you might pursue a particular problem or present an answer. I encourage you to use the blank pages that follow it to make your own observations, notes, drawings, and conclusions. You'll become a teacher yourself through your investigations and writings, so that more such stories can be shared.

The keeping of a nature journal is a time-honored practice, and one directly tied to the very best natural history and nature writing. Your models should be folks like Henry David Thoreau, John Burroughs, and John Muir—patient observers and dedicated note-takers who turned those notes into literature. (More than a century later, their notebooks—especially the pages documenting spring flowering times and insect activities—are being re-examined in light of climate change.)

To help you get started, there are some excellent books on nature journals and nature writing, which I've included in the endnotes for this chapter. Check out what they say about the details of the process: How to make objective observations, free from bias and anthropocentrism. How to patiently utilize all of your senses—sight, sound, and even smell. How to store and retrieve those observations, properly analyzing what you've found. How to document your research, and make connections between your own observations and those of others. And how to present your findings in a scientifically accurate yet engaging manner—the best way to teach others.

But the task is not daunting. It all starts with you and the skills that you already have. So the teacher in me says, "Just get out and do it."

Golden silk orbweavers, once confined to tropical zones, now commonly appear in central Alabama. Is their northward migration due to climate change? Photograph by W. Mike Howell.

Curtis Hansen searches the Walls of Jericho, Jackson County, Alabama, for rare and unusual lichens. Photo by L. J. Davenport.

chapter one

Bioblitz
The Walls of Jericho

A nature journal must be based on personal observations. And the simplest thing to do, once those observations are made, is to fuse them together to describe a place or tell a story. I wrote the following piece directly from notes scribbled during the Bioblitz, straying very little from the actual chronology of events. (Only the rattlesnake story got switched around a bit, just to increase the drama.) The trail in and out of the Walls of Jericho provides a convenient beginning and end, with Davy Crockett's spirit as an obvious theme. I also wrote this piece to commemorate the wonderful work of Forever Wild, on whose board I proudly sit.

Day one dawned cool and cloudy as I scrambled down the Red Trail from Highway 79 near Hytop in northeast Alabama. And "down" was definitely the operative word, my feet falling fast toward an unseen chasm, then catching themselves just in time to meander back and forth in a seemingly endless series of curves and spirals. Midway down the trail, sandstone gave way to limestone, layered like massive fortifications (against erosion) and pocketed with enormous sinkholes. To fortify myself during the two-mile trek, I whistled the "Davy Crockett" theme song. After all, this was Davy's country, the Hurricane Creek where he loved to hunt and the mysterious Walls of Jericho. Was it here that he "kilt him a b'ar, when he was only three"?

Following respectfully in Davy's footsteps, I hiked these hallowed hollers as part of the Walls of Jericho Bioblitz, an effort by the Alabama State Lands Division to "blitzkrieg" the biology of this Forever Wild tract: 12,510 Alabama acres adjoining 8,943 in Tennessee. (Established by an Alabama constitutional amendment in 1992, the Forever Wild Land Trust spends interest earned from offshore natural gas leases on significant land purchases. In this case, the tract's designated uses include hiking, horseback riding, and limited hunting.) So thirty of us biologists descended (quite literally) on this remote countryside, turning over logs and scrutinizing leaves in a systematic, three-day effort to determine what lives there.

After finally reaching base camp along Hurricane Creek that first day, I wandered up nearby Turkey Creek toward the Walls, a dark-green world broken by shafts of golden sunlight. The creek's ice-cold water cascaded and tumbled before completely disappearing underground, leaving only lonely boulders. I admired the innumerable trilliums and twinleafs, the elegant mix of mosses and succulent sedums, and the bizarre walking fern (whose drawn-out leaf tips take root, thus "walking" the host plant across the landscape). A summer storm struck without warning, thundering down upon me with ferocious splendor, and I wondered, "WWDD (What Would Davy Do)?" So I hunkered down under a convenient rocky overhang and contemplated life's great uncertainties while gnawing on a piece of jerky. Smugly dry, I returned to camp to find most of my belongings drenched due to a minor tent-raising miscalculation. Sorry, Davy!

After supper, the evening evolved into caterpillar hunting, as folks fanned out with flashlights to cautiously inspect the undersides of tree leaves for the often hairy and sometimes dangerous larvae. (During the weekend, the amazing total of over twenty species was secured by this communal effort.) We then fell asleep to the far-off calls of hoot owls and whippoorwills.

Day two dawned crisp and clear. (Later on, the hot sun graciously baked the dampness from my kit.) I sipped my (insipid) instant coffee while admiring a pair of red-shouldered hawks, soaring high above, crying forlornly to each other. Was it important conversation that they shared, or just the basic animal need for companionship?

Two snail experts slowly sifted through mounds of leaf litter, searching for tiny treasures. I opted, instead, to join Auburn University's Curtis Hansen on a return trip to the Walls. Hansen studies lichens, those strange amalgamations of algae and fungi that cling to tree bark and rocks. We followed the muddy, rain-slick trail, stopping frequently so that he might "kiss" those surfaces. (A most sexy lot, lichenologists inspect their intended prey so closely that they appear to be smooching.) Finally we reached the Walls themselves, smooth flat faces rising one hundred feet straight up. Hansen carefully scraped off lichen-encrusted pieces with his hammer before we skidded downslope to the creek. (In a most ignominious display of un-Davy-like grace, I slipped slap-down on my, uh, backside.) Here the creek fanned out into a broad series of pools and waterfalls. Continuing upstream and around a sharp bend, we suddenly entered an amphitheater-like bowl, the worn and etched limestone mottled with gray-pink lichens, the shallow solution cavities teeming with darting black tadpoles. Water gushed from caves along the bowl's edges, and Hansen pointed out the aquatic lichens clinging to the wet channels. (*I never know'd of sech!*) I left my companion kissing his new friends and climbed up and over the dry falls. A fearsome rustling spooked me. Was it some hideous, hungry varmint?! No, just another snail person gathering leaves.

That afternoon I joined a small party to wade across Hurricane Creek and explore Polly Anne Spring. Truly fetching and serene, the spring's cold, clear water shimmered over emerald green liverworts and moss-clad cobbles. But on the return trip we just about stepped on a rattlesnake—coiled but (gratefully) asleep. WWDD? While my comrades closely inspected the serpent, debated its taxonomy (Canebrake? Timber rattler?), and fabricated yarns about its length (Five feet? Six?), I skedaddled back to camp.

Then a most efficient crew of aquatic biologists marched in, donned their wetsuits, and scoured the creek bottom with snorkels and seines, finding six species of mussels. And the master list of critters swelled: thirty-one fishes, twenty-eight dragonflies and damselflies, twenty butterflies, fifty-one birds, more than one hundred mosses, more than eighty lichens, twenty-four amphibians and reptiles (including the rattlesnake).

Day three started too early, as I waked to the predawn avian chorus.

(Songbirds sing in the dark when their predators can't see them.) I set out alone, except for a list of local plants, and checked off every species that I spied. ("Raised in the woods so's he knew every tree.") I smiled at the *rat-a-tat-tat* of two pileated woodpeckers signaling each other, slamming their beaks into dead tree limbs, then raising triumphant, raucous squawks. I played cat-and-mouse with a catbird, zeroing in on its mournful, mewing cry in a vain attempt to spot its blue-gray plumage. An indignant indigo bunting whizzed past my left ear, then irritably chirped from a nearby shrub. Had I ventured too close to his mate and nest?

That afternoon, I bid farewell to the snail people (still patiently sifting), the Walls, and Hurricane Creek, and trudged back up the Red Trail. (The reported "world record" for hiking out is twenty-four minutes, but I didn't bother to attempt it.) Two heart-pounding hours later, the highway sounds alerted me to the approaching trail's end, and I turned for one last glimpse of the forest primeval. Davy Crockett, the King of the Wild Frontier, would be proud. Not of me, but just knowing that we've preserved one of his favorite places, and all of its incredible diversity, forever. And wild.

Choose a path through the woods. As you slowly walk along, take careful notes on what you see and hear. What wildflowers appear? Is each one restricted to a single type of habitat? Describe one wildflower so completely that you can use that description to later identify it in a nature guide. What birds inhabit this area? Again, describe one so completely that you can later identify it. Can you differentiate one bird's song from another's? Describe a bird's song in such a way that you could teach it to someone else.

A male soldier fish, or rainbow darter, patrols the riffles and pools of Copper-
run Branch in Limestone County, Alabama. Photo by W. Mike Howell.

The Soldier Fish

W. Mike Howell and I were seining a small creek in Limestone County, Alabama, when up popped a soldier fish. And as we stood in our waders in the knee-deep water, Dr. Howell regaled me with the story of the naming of this fish. And I memorized that story on the spot, knowing that I had to tell it to my readers. I had some fun (and took some liberties) with the soldier motif, of course.

Legend has it that on the eve of a particular Civil War battle, one army stared nervously across a narrow creek at the other. Obviously, the morrow promised a cataclysmic confrontation, with many lives lost and, perhaps, the war's tide turning. Should they risk it all and attack? Would victory be assured? So, in the best time-honored fashion, the officers kneeled and prayed (very loudly and very long), invoking the Almighty to *please* show them a sign that they would, indeed, successfully smite their foes. Soon after, a private crept down to that creek and returned with a much-needed bucket of water. And in it wriggled the sign they had sought.

As Confederates told it, the bucket contained a single fish, just three inches long, but the likes of which no one had ever seen. Brilliant reds and blues splotched its body and head, with similarly hued bands on its dorsal fins. In addition, the blue-green anal fin sported a central red-orange star, while nine rectangular bars decorated its sides. To the prayerful army, the creature's Stars and Bars obviously predicted TOTAL TRIUMPH! (Of course, in the Yankee version of the story—which likewise ends victoriously—the fish exhibited only the Union red, white, and blue.)

Although brand-new to the above combatants, the "soldier fish" (now more commonly called a rainbow darter) is well known to ichthyologists. Like other darters, this species lacks an air bladder to aid flotation, so it scoots or darts along stream bottoms in short, erratic bursts, attacking and dispatching prey before encamping (albeit briefly) on the bottom again. But unlike other darters, which involve a few troops deployed to a single watershed, rainbows fan out broadly (and abundantly) in the Great Lakes and Mississippi and Tennessee river systems, ruthlessly driving out all competitors in pursuit of total Darter Domination.

Soldier fish typically command gravel- or cobble-bottomed streams with numerous riffles and runs. There, juveniles post a wary watch on their elders, with the latter bivouacking in deep, swift riffles while the former bunk in quieter areas near the margins of runs or pools. Rations include a variety of insect larvae, especially mayflies, blackflies, caddis flies, and midges. David Starr Jordan (1851–1931), the nestor of North American ichthyology (and former president of Stanford University), captured the habits and personality of this "gaudiest of all freshwater fishes" thusly: "The Rainbow Darter is a chubby little fish, as compared with other Darters. In its movement it is awkward and ungraceful, though swift and savage as a pike. One of the mildest of its tricks . . . is this: It [will] gently put its head over a stone and catch a water boatman ['walking' on top of the water] by one of its swimming legs, release it, catch it again, and again release it, until at last the boatman, evidently much annoyed, [swims] away out of its reach."

Ah, fun with food!

Mating occurs from late spring to early summer. A male—resplendent in his best dress uniform—stakes out and defends his riffle against all invaders, threatening would-be rivals with vicious fin-to-fin combat. (And size *does* matter: the larger the male, the greater the intimidation.) A female—marked by drabber colors and a lascivious smile—cautiously approaches from downstream, then buries her ventral side in the substrate at the foot of that riffle, where her soldier suitor promptly fertilizes three to seven eggs. Forming a shifting, protective guard around his beloved, the male marches her a short distance upstream where she repeats her half self-burial. The happy couple continues this exhausting regimen over and over until about eight hundred eggs are deposited. (*At ease! Take five! Smoke 'em if you got 'em!*) The tiny, unprotected eggs hatch in ten to twelve days, while the larval stages last another fifty, by the end of which the juveniles have grown to fifteen millimeters long— that is, unless a marauding muskie or pernicious pike gobbles them up. And by their first anniversary of "enlistment," male recruits vigorously defend the breeding riffles, thrashing smaller comrades (and servicing any females) who dare to enter.

That's the story of the soldier fish. Now, I'm not sure which side prayed—and which side triumphed—in that legendary battle. But the fish remains, patrolling clear, gravelly streams of the Middle States and assaulting all intruders. Not just an omen of war-time victory, but a warrior all its own.

Join a nature walk led by experts. Listen as they describe the natural world and name the organisms that are seen. Choose an organism with a unique or intriguing name and research the history of that name. Does its name help to tell the organism's story?

A great blue heron perches on a pine bough along the Gulf Coast.
Photo by W. Mike Howell.

Great Blues

*I grew up in the Pacific Northwest, where the Native Americans
are famous for their totem poles. And the idea of a totem animal—
a creature to guide and assist you through life—has always been a
powerful one for me. I wrote the following piece in homage to my
own totem animal, the great blue heron. But I also wrote it to pre-
serve a childhood memory and to capture my father's personality—
a secondary task, but equally important.*

Some childhood memories haunt us forever, evoking powerful feelings that refuse to dissipate with time. For me,
one such memory involves a great blue heron.

Way back when, my family enjoyed a little cabin on a
little lake near a little town in western Washington. Sharing it with us was a lone "great blue." Dapper as a Colonial
colonel, he sported a gray-blue tailcoat with black epau-
lets, chestnut cravat and matching knickers, and plumed
black-and-white hat. Four feet tall (towering over me at
the time), he patiently patrolled the lily pads and reeds
along the shore, dignified and unflappable—until disturbed by curious tykes, when he became quite flappable!

Then, uttering an irritated "kraak" and utilizing all six feet of wingspan, he stroked slowly and powerfully to the opposite shore, legs hanging below like rudders, neck hunched back like a flying Ichabod (Heron) Crane.

The largest, most widespread, and best-known wetland birds of North America, great blues frequent fresh- and saltwater shorelines from southern Alaska to Mexico, coast to coast. (To celebrate this ubiquity, both parts of their scientific name, *Ardea herodias,* mean "heron": first in Latin and then in Greek.) Pacific Northwest populations never migrate, while the rest overwinter in such exotic locales as the Caribbean, Greater Antilles, and Venezuela. Supreme opportunists, they forage for fish, crayfish, snakes, frogs, salamanders, and anything else foolish enough to venture close. The primary hunting mode, Standing Still (not to be confused with the more energized secondary mode, Walking Slowly), requires the hunter to remain completely motionless, neck curled back in the same S-shape used in flying; while the head never moves, the eyes change focus constantly, backward and forward, left and right. Then, swiftly and violently, its neck straightens, hurling its head into the water, snatching up prey in a serrated, pincer-like bill. The hapless victim is perfunctorily tossed into the air and swallowed headfirst; the waiting game begins anew.

Solitary creatures, great blues congregate only for migration, mating, and raising their young. Males return first to the nesting sites, "heronries" or "rookeries" far from encroaching civilization—the edge of a lake or swamp, with a stagnant moat to thwart land-based predators. There, at the ends of stout branches, they weave together pencil-thin sticks to construct shallow platforms, each three feet across. Females soon follow and, after a short courtship marked by "frenzied exhibitionism" of their suitors, three to seven greenish-blue duck-sized eggs are laid. For the required thirty days of incubation and sixty days of feeding, the parents take turns at the nest, greeting each other with a guttural "ar-ar-ar." ("How ar-ar-ar you? Ar-ar-ar you mad that I was gone so long?") Despite the adults' diligence, only half of the initial eggs finally fledge; with staggered hatching, older nestlings grab the lion's share of the proffered food or push little brothers and sisters out and down to their demise. And only a third of all fledglings survive to reproduce; poor hunting skills doom most to starvation.

While these threats are natural, others come directly from us. One hundred years ago, herons, cranes, and egrets were literally under fire— hundreds of thousands of large, slow-flying, tempting targets, their feathers in great demand by the millinery trade. (No wonder great herons squawk the blues!) While federal laws and international treaties now minimize this threat, population numbers still suffer from wetland loss, development pressure on rookeries, and water pollution, especially by paper mills.

But back to that poignant childhood memory. One particular Saturday, an explosion shattered the morning's gentle routine; much excited shouting soon followed, and my dad and I hurried over to a neighboring cabin where the commotion originated. Standing on his dock, our neighbor waved a shotgun and pointed proudly to the water below. There floated our avian friend, his wings oddly bent, his neck grotesquely twisted, his beak strangely agape, his eyes vacantly staring. When asked, our neighbor boasted that he shot the bird "because it was eating too many fish!" My dad, never the confrontational type, said nothing. But I sensed his anger and shared his sadness at this pointless death. We walked home in silence, each coming to grips with the day's events—the first time I had ever heard a gunshot (except on TV westerns), the first time I had ever beheld death, the first time I had ever felt the awful weight of injustice.

I think about that great blue heron every time I spy one of his surviving cousins. And, for very personal reasons, I take a moment to thank him. Both in life and in death, he inspired a small boy to wonder and worry about nature—its beauty, its intricacy, its drama, its joy, and (too often when humans are involved) its sorrow.

Do you have a totem animal, a wild creature (not a pet!) that relates to you more than any other? Which of its special characteristics are so attractive to you, and why do you feel this unique affinity? Do you share a childhood experience, perhaps? Write about that animal and the relationship you share, capturing that personal memory.

A freshly dug-up doodlebug, half an inch long, brandishes its menacing pincers in Jefferson County, Alabama. Photo by W. Mike Howell.

chapter four

Doodlebugs and Ant Lions

This piece is also based on a childhood memory, but far less person-
al than the previous one. I've long been intrigued by the universal
experiences that children enjoy outdoors, no matter where their
particular outdoors happen to be.

You knew them first as children, crawling under the back
porch of Grandma's house to peer into those mysterious
dimples in the warm sand. Perhaps with a pine needle, or a
single hair purposely plucked from your sister's scalp, you
poked at the slopes of a cavity, the sand grains cascading
abruptly toward the bottom. And maybe you whispered a
brief incantation:

> Doodlebug, doodlebug,
> Come out of your hole.
> Your house is on fire,
> Your children are alone.

Then a flash of movement, and the barest glimpse of a pair
of pincers, and your playmate disappeared.

Two thousand species of doodlebugs dimple the Earth's

surface, so that children everywhere share such close encounters. The larvae of primitive insects called neuropterans, doodlebugs metamorphose into damselfly-looking adults inhabiting sandy-soiled areas, laying their eggs at dry cave openings (or under porches and eaves) or on open forest floors (or abandoned city lots). Once hatched out, the larvae wriggle their ways across the vacant surface, leaving meandering trails resembling the musings of preoccupied artists.

A doodlebug doodles until it finds a perfect hideout—a sunny spot sheltered from wind and rain. Pushing itself backward, a 'bug draws a quarter-sized circle on the ground, then digs deeper and deeper, spiraling toward the center, flinging out granules with its head. After fifteen minutes of such neck-jerking work, it buries itself at the bottom of the funnel-shaped chasm with only its head and open jaws exposed. Thus, a whimsically doodling 'bug transforms into its evil twin, the ferocious and aptly named ant lion, deep in its Dimple of Death. And there it waits.

But First, Two Fun Facts: (1) The size of the hole is not related to the size of the 'bug but to its hunger, with the hungriest critters crafting the biggest craters. (2) 'Bugs dig bigger pits at the full moon. Even in captivity (as in that sand-filled aquarium on your childhood dresser), or without lunar cues at all, they'll stick to a 29.5 day menstrual cycle.

Soon an oblivious ant wanders too close to the edge of an ant lion's trap. After sliding halfway down the slippery slope, it frantically scratches and claws its way back up. In response, the ant lion hurls a grain of sand, causing the captive to lose its grip and fall straight into the huge (to the ant), curved, piercing mandibles. The predator immediately seizes its prey, injects it with paralyzing poison, and sucks out its vital juices. Once satiated, the lion uses the same throwing motion of its head to toss the lifeless husk from its lair.

Such ant/ant lion antics continue up to three years, when the insect burrows deeper in the ground, extruding white silken threads to form a spherical cocoon. There it metamorphoses into a pupal stage; three weeks later, a tiny-winged adult eases out through a small hatch, clambers to the surface, and climbs up the nearest plant. (Which is why, in the dimpled Zen Garden of Death on that childhood dresser, you *should have* stuck at least one twig.) After twenty minutes of wing expansion and hardening, the adult's ready to fly. Adulthood—during which a mate must be procured and eggs must be deposited—lasts less than a month.

Now, Two Not-So-Fun Facts: (1) An ant lion's digestive tract has no "bottom" opening, so that all those Ant McNuggets produce three years' constipation. *Is that what makes ant lions so mean?* (2) Adults often lay their eggs too close to active pits, and so are occasionally captured and devoured by their younger relatives.

The contrasting views of companionable doodlebugs and voracious ant lions appear often in American literature (and even in our history). In *Cape Cod,* Henry David Thoreau uses ant lion imagery to describe a desolate hamlet: "We began to think that we might tumble into a village before we were aware of it, as into an ant lion's hole, and be drawn into the sand irrecoverably." Early in John Steinbeck's *The Pearl,* the hapless Kino—in a not-so-subtle bit of foreshadowing—watches "with the detachment of God" while an ant futilely struggles to escape its fate in an ant lion's den. In contrast, the hero of Mark Twain's *The Adventures of Tom Sawyer* desperately seeks counsel from an insect oracle: "Doodlebug, Doodlebug, Tell me what I want to know!" (Because the oracle "dasn't tell," Tom reasons that a witch must have caused his current predicament.) And Apollo 16 astronaut Charles Duke, while gazing into an enormous lunar crater, recited a favorite childhood charm: "Doodlebug, Doodlebug, Are you at home?" Fortunately for us (and Mission Control), he received no discernible answer.

The global, universal nature of your childhood incantations has been examined, too. Anywhere on Earth, and in any language, doodlebug/ant lion charms contain the same basic elements: (1) a beseeching to come out, (2) an enticement (usually of a local food or beverage) to come out, and (3) some dire consequences for *not* coming out. So in their own weird way, doodlebugs and ant lions unite the children of the world—through mystery, secret friendships, folklore, tribal chants, and certain (for the ant) violent death.

Again, think back to your childhood. How did you experience nature as a child? What did you learn from that perspective, and what did you conclude? Is there some simple activity, like climbing a tree or poking a stick into a mysterious hole, that you remember with a child's wonder and happiness? Write about that experience—what you saw, felt, or smelled.

An ever-hopeful male fence lizard struts his best stuff in Bibb County, Alabama. Photo by L. J. Davenport.

Fence Lizards

I was walking down a trail along the Cahaba River in Bibb County, Alabama, one sunny afternoon when I stopped dead in my tracks. A lizard sat on top of a small log in the middle of that trail, bobbing up and down, completely oblivious to my presence— in fact, he had no interest in me whatsoever. And as he bobbed, he revealed his gorgeous light blue belly and neck to the world. I had no idea what kind of lizard he was, or why he was doing what he was doing.

This chapter resulted from my identifying that lizard and researching his story. I wrote it, obviously, from a human male's perspective.

It's not easy being male. First you've got to woo your beloved, hoping to captivate her with your looks, charm, and physical prowess while intimidating all rival suitors. Then you anxiously await her decision—*yes, no,* or *maybe*—with a good chance to suffer the cruelest rejection. An agonizing ritual we human males share with fence lizards. . . .

Called the rusty-backed, scaly, or pine lizard in many parts of the South, these husky, rough-scaled reptiles are conspicuous both due to their size (four to seven inches long at adult-hood) and tendency to bask in exposed places. Dark, undulating cross-bands, often in the shape of irregular Vs, mark their rusty- or gray-brown backs. In addition, adult males sport pastel blue sides to their bellies, the bright color bordered by black toward the center of the body. The same combination appears on the throat or dewlap.

Fence lizards populate forested regions, especially piney woods and cut-over woodlands dotted with brush heaps and fallen trees. Logs, stumps, tree trunks, old lumber piles, and split-rail fences are favorite haunts, the creatures blending in perfect camouflaged harmony. When disturbed, they escape by scurrying up the side of the tree or fence op-posite their pursuer, speedily climbing a short distance, then remaining perfectly motionless. They can be easily caught, though, if you practice the Fine Art of Fence Lizard Catching—a slow, cautious approach from in front, followed by a lightning-quick grab from behind.

Dry, open, abused land produces abundant invertebrate life, and fence lizards happily consume all they can—insects, spiders, millipedes, snails, and especially ants and beetles. Movement draws their attention, and only moving prey is captured, usually by a sudden dash or lunge. Most hunting occurs early in the morning, after a revitalizing dawn sun bath; they may forage again toward evening before retiring to shelter for the night.

Being ectotherms, fence lizards also seek winter shelter, hibernating in burrows, spaces under or between rocks, or within rotten stumps or logs. Mating behavior begins soon after their spring emergence. A typi-cal male is strongly territorial, choosing an obvious site and advertising his presence to other members of the species with an elaborate series of shuddering head-bobs and vigorous push-ups, his body moving repeatedly (and perhaps suggestively) up and down. He compresses himself laterally, providing maximal display of his beautiful blue patches; orienting himself at right angles to any intruder adds to the patches' visibility. (Such behavior will be directed toward either sex, especially during initial encounter stages.) If another male enters his territory, he throws down the saurian gauntlet—rapid-fire push-ups plus lateral compression plus ventral extension of the dewlap, all at once. Physical attack ensues, with the first male charging the other and attempting to bite him and chase him away.

If, however, a fetching female appears, and she accepts his advances, then mating occurs. (How could a female, of any species, resist a guy sporting belly and throat patches best described as "leisure suit blue"?!) After six weeks' gestation, she lays a clutch of three to thirteen eggs—each egg one-half inch long with a thin, papery shell—in a shallow cavity, deep enough to unite the constant moisture of the soil with warmth conducted from the sunlit surface. The hatchlings, about one and a half inches long, emerge two months later.

But what if she's less than impressed by his performance? Females being courted are quite likely to respond with—*it hurts me to even put these words together*—a rejection display. She presents her own lateral view to the male, revealing the lack of belly patches and clearly identifying her gender. ("Yes, I'm a female, *but I just don't want you.*") She arches her back, performs an energetic feminine version of Shuddering Push-Ups, then quickly sidle-hops away, out of his territory (and life) forever.

And at this point, the comparisons with human situations have become too personal, and painful, for me to continue.

On your next nature walk, concentrate on watching and recording the behaviors of the animals that you see, especially reptiles and birds. How do they hide from predators—through behavior, camouflage, or both? In contrast, how do they reveal themselves to potential mates? Do the males take more chances than the females?

_____

_____

_____

_____

_____

_____

_____

_____

_____

_____

_____

_____

_____

_____

_____

_____

_____

_____

_____

_____

An American rubyspot male (*left*) and female form the wheel position in Bibb County, Alabama. Photo by L. J. Davenport.

The Wheel of Life

This piece is linked to the previous one in that it's based on an observation along the same stretch of the Cahaba River—again, something that I saw but didn't immediately understand. Fortunately, this time I had Dr. Randy Haddock of the Cahaba River Society with me, and he very plainly and accurately described the wonderful wheel formation of damselflies.

Snake doctors, devil's darning needles, horse stingers, and mosquito hawks—these colorful names reflect the folklore and superstition of dragonflies and damselflies. Winged emblems of victory in Japan, their less-esteemed American cousins purportedly sew children's ears together and resuscitate dead snakes.

Biologists place the two closely related groups—ferocious, demonic dragons and refined, dainty damsels—in the order Odonata; the name refers to the tooth-like ends of their abdomens. (Sorry—no stingers or needles!) Fleeting streaks of iridescence, these slender cylinders sport tasteful (and perhaps tasty) pastels, harsh metallic tints, or gaudy stripes and blazes. Two pairs of wings—perfectly equal in damselflies, not quite so in dragonflies—adorn the thorax. (Thanks to independent control of each appendage, the bearer can brake sharply, spin quickly, and even fly backwards.) At rest, the latter group keeps its wings perpendicular, ready for instant flight, while the former folds them neatly along the back.

Damsels and dragons depend on water—in fact, they spend the majority of their lives as darkly camouflaged aquatic larvae. Skulking in the shallow bottoms of streams, ponds, lakes, and marshes, these insatiable nymphs or naiads bushwhack and dispatch an assortment of hapless critters—including brother/sister odonates and even small fish—thanks to a monstrous distensible lower jaw. This labium shoots out half the naiad's body length, skewering a victim on pointy prongs before yanking it back toward churning teeth. In turn, unlucky little larvae serve as tasty snacks for bigger brethren.

To respire underwater, damselfly larvae rely on a trio of caudal gills, fanning them out into the water column to gather much-needed dissolved oxygen. In contrast, dragonflies respire rectally, sucking in fresh water through the anus and—just like a rubber bulb syringe—expelling it by the same route. More than just good clean fun, the resulting jet propulsion hurtles a nymph toward waiting prey or scoots it away from danger.

After many months (and molts), an odonate adult-to-be swims warily to the water surface, inspiring air for the first time while crawling up a stem or rock. As internal pressure increases, the larval "skin" or exoskeleton cracks open; the new adult slowly and eerily rises, a weak and vulnerable ghost. Resting briefly to harden its body and inflate its wings, the imago soon flies away, leaving a lifeless case (*exuvium*) still tightly gripping the substrate.

On this maiden flight, and while fully maturing, adult dragonflies may venture far from their watery birthplaces; damselflies stay much closer. But they both return, cruising the stream bank or pond margin, entangling mosquitoes, mayflies, and smaller kinfolk in their spiny-legged "baskets" before savagely devouring them. Miniature hawks or eagles, odonates enjoy superior sight, with huge, protruding, goggle-like eyes taking up much of their heads. (A housefly has 4,000 units making up one compound eye; a dragonfly may have seven times as many.) And those heads swivel readily on a slender neck, engineering motion perception unparalleled in the insect world.

Eyesight forms the basis for odonate sex—in big words, "visually mediated reproductive behavior." A male stakes out a suitable territory and—goggles securely in place—fights *Snoopy vs. Red Baron* aerial battles to drive all competitors away. Anticipating that a willing partner will wander close, he curls his abdomen downward and forward, transferring sperm from true genitalia on segment 9 to a pouch containing secondary genitalia on the underside of segment 2. (*That's right, guys—he mates with himself, and has two sets of parts!*) Hovering over a hovering female, he cradles her head and thorax with his legs, then grabs the back of her neck with anal claspers. Firmly connected, they fly off in tandem like two tiny airplanes, one being refueled by the other. But no "fuel" transfers until the bride, safe on a perch, imitates the groom's previous contortion, looping her abdomen downward and forward to connect to his accessory genitalia, completing the wheel position. A single such mating fertilizes several hundred eggs.

More heart-shaped than circular, the Wheel of Love may last a few minutes or several hours. Then, back flying in tandem, the male gently guides his beloved to a pre-selected site within his watery domain. In some species, she "bombs" the rippled surface from the air; in others, using a knife-like ovipositor, she carves holes in aerial plant stems or soft wood before inserting the eggs; and in others, she descends completely underwater—for up to two hours—clutching a bubble of reserve air between her wings. The male stays close, guarding her from rivals, hoping to "get lucky" once more upon her wet and bedraggled ascent. (The Wheel of Fortune?) In a few weeks, the next generation of nymphal ogres emerges.

Goggle eyes, jet propulsion, distensible jaws, and secondary penises— there's far more than nicknames and folklore to the damsel and dragon story. And these incredible innovations make the Wheel of Life go 'round.

Again, use a nature walk to observe the behaviors of animals, but this time, concentrate on insects. (Grasshoppers and ants might be good choices.) How do they communicate with each other—through sight, sound, or touch? Do members of the same species congregate in groups, or do all individuals remain separate?

A giant swallowtail sups on a salvia in Baldwin County, Alabama.
Photo by W. Mike Howell.

chapter seven

Giant Swallowtails and Metamorphosis

I wrote this piece after a field trip to eastern Texas, where I inspected and admired a very compliant giant swallowtail. As I researched the subject, I became convinced that insect metamorphosis is TOTALLY AND UNDENIABLY AMAZING. I also became obsessed with "animal taxonomy" or how animals have their own means to identify the plants needed in their life cycles. How does a giant swallowtail "know" a member of the citrus family? What characteristics does it use to make the correct choice of a host plant for its eggs?

Butterflies are the stuff that myths are made of. With their flashing colors, elusive flight, and amazing life history, butterflies have long held a treasured place in human lives, considered to be messengers from the gods and symbols for the immortal soul.

One of the most spectacular of these aerial messengers is the giant swallowtail, *Papilio cresphontes*. Ranging from

Canada to the Mexican border, "giants" are the largest North American butterflies, with some individuals measuring six inches from one black-and-yellow wing tip to the other. They are most frequently found in open woodlands, pastures, and orchards, the larvae feeding on citrus trees and prickly ash.

The giant swallowtails that we enjoy, though, comprise just one short page of a complex story. Like other butterflies, giant swallowtails undergo complete, four-stage metamorphosis. The adult females, soon after mating, lay yellow or light green eggs on host plant leaves and twigs. Hatching after a few days, the larvae or caterpillars remain on the host trees, gobbling up tender new shoots and leaves. (Colloquially called "orange dogs," they may become major pests in citrus groves.) To escape detection, the larvae are camouflaged to resemble bird dung—mottled brown with a cream-colored saddle on top. They also possess an orange-colored *osmeterium,* an eversible forked organ just behind the head which emits a powerful, repugnant odor when extruded, as when disturbed by a predator.

After a series of molts, larvae enter the pupa or chrysalis stage of development, attaching themselves to a vertical surface while building a hard outer case; like the harness of a telephone lineman, a silk strand or girdle secures the middle of the body. In a process called histolysis, most of the larval organs dissolve, and the resulting fluid reforms ("histogenesis") about small primordia. The enormous digestive tract—so essential to the life of the larva—is drastically reduced, while the eyes, brain, sex organs, and flight muscles are greatly enhanced. After completing these transformations, the adult emerges, inflates its wings with hydraulic fluid, and flies off to find a mate and repeat the cycle. Three of these cycles will be completed during a typical long, hot summer.

From a biological standpoint, metamorphosis allows for the larva and adult to live completely different, non-competing lives. A homely and awkward caterpillar, dominated by its need to feed, is transformed into a beautiful and graceful adult, dominated by its need to mate; an "orange dog," restricted to a single host tree, becomes a giant swallowtail, free to float the breeze.

But in symbolic terms, the process means much more. The wormlike caterpillar enters a deathlike pupal state, only to be resurrected as an ethereal-winged adult floating skyward. Is it any wonder that the ancient Greeks used the same word, *psyche,* for both the butterfly and the soul?

Choose a local butterfly and research its life history, especially its host plants. (How are the host plants related, botanically speaking?) Then find one in nature and follow it. (You might want to use binoculars.) Does it utilize the plants listed in the literature or are there exceptions? Can you find the tiny eggs that the female secures to the back sides of host leaves? (A hand lens would be useful.) After locating the eggs, keep coming back to that plant to monitor the hatching, feeding, and growth of the larvae. How destructive are the caterpillars in their feeding?

A newly emerged luna moth clings to a sweetgum trunk in suburban Birmingham, Alabama. Photo by W. Mike Howell.

chapter eight

Luna Moths and Pheromones

This chapter consists of two parts: the moth life cycle and the pheromone phenomenon. While I started off interested mainly in the first part, I found myself drawn more and more into the broader-reaching aspects of the second.

Few summer sights match that of a male luna moth hovering near a lamp post, dodging and darting, zooming to and fro, desperately seeking something that he can't quite find. And in his frenzied search he teaches us a little about ourselves.

Spectacular, unmistakable, fully four inches across, our luna's pale green wings sport transparent eyespots rimmed with yellow and maroon; as an extra touch of elegance, the bottom wings flare out into delicately curved tails. Dark pink legs, orange antennae, and a finger-thick, furry white abdomen complete the handsome portrait. Found only in North America, lunas frequent deciduous forests east of the Great Plains, from the Canadian border to the Gulf

Coast. In northern reaches, where only one brood develops each year, an intense blue-green coloration dominates; in the South, successive sets of offspring are duller yellowish-green.

Preferred food plants likewise differ latitudinally, the caterpillars—veritable eating machines—devouring white birch at one extreme and hickory, walnut, sumac, persimmon, and sweetgum farther south. Plump and pea green, with impressive yellow stripes and six rows of red knobs, the voracious (and to many birds, delicious) larvae anchor themselves with "extra" legs (*prolegs*) equipped with suckerlike pads, twisting and stretching for that next scrumptious bite. Mosquito-sized at birth, they embark on a month-long feeding binge that increases their weight 4,000-fold, stopping only long enough to molt four times.

Upon reaching final size (about three inches long), each caterpillar spins a papery brown cocoon inside a loose wrapping of leaves, either down in the leaf litter or up in the host tree. (Leaves, of course, fall in the fall, joining the cocoon to the litter below.) Weeks later, the adult-to-be noisily heaves himself against one end of his chamber, oozing an enzymatic secretion called *cocoonase* to loosen the silk fibers and then ripping them apart with hornlike projections near the base of each forewing. An exhausting five minutes later, our lepidopteran hero finally frees himself to clamber up a nearby tree trunk and pump fluid into his soft, stubby wings. Late that night, he flutters off on his maiden voyage—most likely in search of a maiden. . . .

In contrast to their larvae, adult luna moths do not feed; in fact, they have no digestive tracts with which to do so, those parts having disintegrated during pupation. Instead, they devote their one week of life entirely to sex, their sole mission to mate. Since adults emerge far apart, a female—usually before her first flight—sends out a chemical "call" to potential mates, extending a gland from the posterior of her abdomen and releasing an odorless, airborne pheromone. It's as if she's saying, *"HEY BIG FELLA! Don't spend another night alone! Drop whatever you're doing and get over here—PRONTO!"* And, seeing as how he has no prior commitments and his life has this sole purpose, he immediately obeys. Using exquisitely tuned receptors at the tips of his feathery antennae, he doggedly tracks her scentless trail through miles of uncharted wilderness, determined to make full use of this *one chance* at reproduction— although he may be confused by a street lamp or (sadly and lethally) suckered in by a backyard bug zapper. If his frantic search ends with successful coupling, the seductress spends her remaining lifespan laying small clusters of eggs on the undersides of host plant leaves. Her lover may follow other trails over the next few nights, but he also soon dies.

When pheromones (and their extreme potency) were discovered in moths fifty years ago, we humans treated them as mere curiosities of the invertebrate world, smugly declaring ourselves to be "above" such crude chemical control. But science has steadily chipped away at that smugness. We now know that queen bees emit compounds that keep would-be successors from ever doing so, while "scents" from female goldfish cause neighboring sperm counts to quintuple overnight. Boar slobber contains androstenone, powerful enough that one whiff brings sows literally to their knees, assuming the mating position. And since human sweat produces the same stuff, a number of somewhat questionable Internet products incorporate this compound— *"Don't spend another night alone! SEND YOUR MONEY NOW!*" (A man using such a product, though, should probably keep his distance from swine of either gender.)

"Okay," you scoff, "but what's *really true* about humans?" Women often note that their menstrual cycles fall into sync, especially when living in close quarters, as on a sports team or in a college dormitory. A 1998 study—first done with rats and then with humans—shows that two pheromones work together, one released before ovulation that speeds up another woman's cycle, and one released during ovulation that slows it down. And women who spend more "quality time" with men have shorter and more regular cycles. (It's *conceivable* that such findings may prove helpful in treating infertility!) And, in a way eerily similar to that in goldfish, the vaginal tract produces secretions (cleverly called *copulins*) that elevate testosterone levels, boosting a receptive man's libido and, in classic medicalese, "positively affecting perceptions of female attractiveness in targeted males." In other words, *actual scientific proof* that everyone looks better near closing time!

So, like the luna moths we greatly admire, we humans are directly controlled by the invisible chemical commands that surround us. And we are led around by our noses far more than we realize—or care to admit.

Examine this statement: "We humans are biological creatures, no more and no less, operating under the same laws of nature as other closely related animals." Now examine the human world around you, looking for evidence to support (or refute) this statement. How do we humans separate ourselves from nature, and how are we invariably drawn back? What basic biological needs must we fill every day? Which biological forces can we escape, and which ones dominate us?

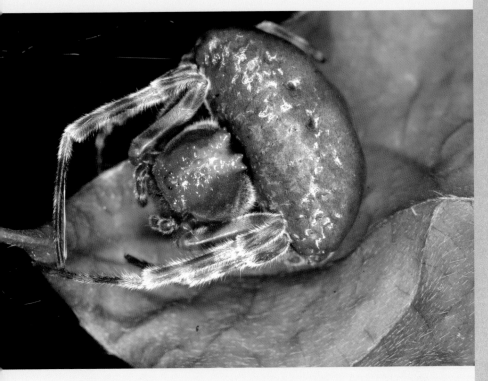

A female bolas spider sleeps on a leaf, resting up for the night's deception.
Photo by W. Mike Howell.

Bolas Spiders

This chapter wrote itself, or perhaps I should say that Charles Hutchinson "channeled" it through me. It's hard to improve on his superb observations and skill at descriptive nature writing, so I used his work as my basic theme, adding details and updating information as needed.

I'm Charles E. Hutchinson's biggest fan. Now, I never met the guy, nor do I know anything about him. But he was the first to see, describe, and understand something that no one else had: the hunting behavior of a female bolas spider.

A Little Background: The decades surrounding 1900 mark the Golden Age of Nature Study, when sophisticated folks pursued the honorable pastime of observing and recording the natural wonders around them. So Mr. Hutchinson sat outside for days, notebook and sketchpad in hand, carefully watching an odd-looking female spider and noting every nuance of her behavior. She obviously belonged to the orb weaver family—a group well known for its fancy webbery—yet she constructed no web at all.

And during daylight hours she simply didn't move, resting openly on the top side of a leaf, cleverly (and safely) disguised as a fat glob of bird poop.

Now the Good Stuff: At nightfall, his subject got incredibly busy. First, just like Walt Whitman's noiseless, patient spider, she launched forth filaments out of herself, "ever tirelessly speeding them" to form a sturdy line, a few inches across, between two branches. Shifting to the middle of her "trapeze," she then created another thread, this time adding what Hutchinson called "a very small quantity of viscid matter," working and reworking it with her spindly legs until it resembled a sticky ball or pendulum, which she dropped into the free space below. To complete the circus motif, she finished her act by hanging sideways, grasping the horizontal line with two of her left legs, the weighted vertical line looped between her mouth and right foreleg.

Continuing with Hutchinson's elegant and precise prose, recorded in a 1903 article ("A Bolas-Throwing Spider") in *Scientific American:*

> If the writer's description is clear, the reader now perceives the spider holding in its hand, as it were, a line to the lower end of which is attached a globule, the whole forming a most singular and ingenious contrivance designed for a useful purpose. . . . If now the observer is to be rewarded, he will see, by the light of the moon, a large moth approaching, flying slowly along as though searching for something. . . . As the insect comes within [a] carefully measured limit, the spider draws back the bolas-supporting leg and, with a pendulum-like movement, swings it rapidly forward in the direction of the moth. The bait is directed with almost unerring aim and finds

lodgment on some portion of the victim. In nearly every instance, it strikes a wing, a part to which it is probably directed. . . . The moth, finding itself fast, flutters violently in an attempt to free itself, but the assailant drops quickly down from its trapeze and sinks its fangs into a vital part. . . . By reason of the poison injected, the moth is soon paralyzed, after which it is carefully enswathed in bands of silk.

Wow! A spider hanging by two legs, swinging a sticky weight or "bolas" like an Argentine cowboy, ensnaring, paralyzing, and then trussing up her victim (to be eaten later) in a silk straitjacket. There in his own backyard, Hutchinson struck the mother lode of Nature Study. His meticulous, painstaking work and vivid descriptions were commemorated fifty years later by the naming of a newly discovered bolas spider, *Mastophora hutchinsoni.* (Less seriously, a sports-nut-turned-arachnologist christened another species *M. dizzydeani:* "Since the spider's livelihood depends on throwing a ball fast and accurately, it seems appropriate to name it in honor of one of the greatest baseball pitchers of all time.")

But Here's What I Love Most: After accurately describing what he had seen, Hutchinson pondered the Why of it all, and then predicted exactly what later researchers would discover. Why, for example, would any reasonably intelligent, self-respecting moth be attracted to a dangling medallion of doom? Hutchinson concluded his paper by stating that "in view of the limited number of moths ordinarily about . . . and the almost unfailing success of the spider in making a capture during the night," a scentless (to us humans) but "agreeable" odor must attract the prey.

The Prediction Fulfilled: Hutchinson's hypothesis hung in limbo until

1977, when another keen observer added this to the puzzle: *Only male moths* zero in on the bolas, and *only from downwind.* Then ten years later, biochemists proved that bolas spiders produce compounds identical to lepidopteran sex pheromones. *That's right!* Females practice the most dastardly deception ever devised, tricking innocent male moths—just out cruising for a good time and minding their own business—by mimicking the scentless scents that they seek. Spiders swallow and reform the bolas frequently during the night (another of Hutchinson's original observations), thus renewing its viscidity. Why not also change its "flavor" to better fit expected prey species—like "early" pheromones for dusk-flying types and "late" ones for the midnight crowd? And why not make seasonal changes as well, to catch bigger critters as you grow bigger? Such utter evil! (At least from a fellow male's perspective.)

So I salute you, Mr. Hutchinson! You were the first to see it and the first to make sense of it, and then you correctly predicted what other scientists would conclude about it. Good work, sir!

Go to a library and find a dusty volume from the Golden Age of Nature
Study (1880–1920). Take your time while leafing through it. What is
its theme, and how is it organized? Is it illustrated with photographs
or line drawings? Find a treatment for a local plant or animal and read
it carefully. What details of the organism's life cycle are described, and
how? Are the words accurate and precise enough to fully capture that
organism? Do you know of any modern data that would support or
refute the author's treatment?

A pink lady's-slipper orchid in all of its enticing, deceptive glory in DeKalb County, Alabama. Photo by L. J. Davenport.

chapter ten

Pink Moccasins

This chapter is ostensibly about beauty—but it might be about grotesqueness, because few flowers are as grotesquely beautiful, or as perfectly and deceptively constructed, as lady's-slipper orchids.

Ah, beauty! How we marvel at its perceived perfection, briefly ignoring just how deceptive and costly (and entrapping) beauty can be.

Perhaps the most beautifully deceptive, and deceptively beautiful, wildflowers in North America are pink lady's-slipper orchids or moccasin flowers (*Cypripedium acaule*). Aptly named for their oddly shaped lowermost petals, pink moccasins occupy wet piney woods, swampy places, and bog margins. Widespread across northern and central climes, the species reaches its southernmost distribution in northeastern Alabama, where it often grows in huge colonies of hundreds of individuals.

Flowering in pink moccasins occurs from April to July, depending on latitude, with a single bloom arising from between two silvery, pleated leaves. The sepals and lateral petals appear as weird, greenish purple, twisted growths, exuding a sugary sweet aroma; the median petal or labellum is inflated into a two-inch-long shoe-shaped sac with a prominent dorsal cleft. Magenta, rose-pink, or nearly white, with darker veins as "nectar guides," this labellum forms a strangely pulchritudinous—perhaps grotesque—hanging pouch.

But beauty, as we know, is in the compound eyes of the pollinator. And for pink moccasins, those "bug eyes" belong to a bumblebee queen. Enticed by the colors and fragrance, and trusting the veins as her guides, she tumbles between the bulging lips, happily humming in anticipation of nectarous reward. *But wait! What's the deal? Where's the good stuff?* The realization suddenly hits her: She's been duped—no nectar to be found and no easy way out. The inrolled margins of the aperture, along with minute hairs aimed in the same direction, prevent her leaving by her initial route. She must either force her way out through one of the exit canals at the back of the flower, or perish in a most glorious death chamber. (The carcasses of insects that fail to escape this insidious trap will be greedily consumed by crab spiders lurking in its corners.) Fumbling about in the dim light, she first bumps into the stigma, depositing pollen grains from a previously visited flower in the process. Then an anther rears its swollen head, directly blocking her path. Now dusted with sticky pollen, she stumbles out a narrow slit, relieved by her equally narrow escape but grumbling loudly, much wiser for the experience.

Okay, make that *slightly* wiser. Deceit-based pollination—where all signs point to a sweet treat, but none is produced—requires naïve, inexperienced, just-emerged-from-hibernation, still-wet-behind-the-antennae bees. Such creatures abound in springtime, buzzing about the countryside, exploring fresh habitats and sampling different flowers. And many are optimistic and forgiving (and foolish) enough to sample pink moccasins twice or more in succession, thus effecting pollination.

However, bees are not stupid—or, at least, not for long. Their preprogrammed and interdependent lives reflect sound economic principles—a compensating amount of sugary energy must be gained for every unit spent. (Bernd Heinrich, combining ecological data with basic economic analysis in his book *Bumblebee Economics,* proves this quite elegantly.) Once bumblebees wise up to the trick being played, they quit playing and turn their attention to more profitable pursuits. Statistical studies show that, despite the huge number of flowers in a given area, very little pink moccasin pollen is ever transferred, resulting in very poor fruit set—barely 10 percent in many populations.

But let's not feel sorry for lonely lady's slippers, prisoners of a system leading to their inexorable extinction. Following the same economic and energy rules that govern their pollinators, each of those few fruits produced contains ten thousand to fifty thousand tiny seeds—lots of results for little expenditure, and more than enough to maintain the species (and its treachery).

So, in the end, do beauty and deception always win?

Select a wildflower and examine it closely in the field. Identify the different parts: sepals, petals, stamens, and pistils. What purpose does each part serve in reproduction? How does your flower accommodate its pollinator—a long throat for a long tongue or a landing platform for a heavy insect? Based on the flower's structure, what prediction (hypothesis) can you make as to its pollinator? Do your field observations support that hypothesis?

The common woodland plant, jack-in-the-pulpit. Or is it a "jill"?
Photo by W. Mike Howell.

chapter eleven

Jack- (or Jill-) in-the-Pulpit

This chapter comes directly from a research project conducted by my student, Carolee Franklin. While it involves another oddly flowered plant, the oddity is mainly in how the plant responds to its environment before, during, and after successful reproduction.

Sex is never simple, nor without its costs. While the high costs of sexual reproduction—attracting a mate, transferring gametes, and supporting the offspring—are obvious in animal species (including humans), the same lessons can be learned from a common woodland plant, jack-in-the-pulpit.

"Jacks" (*Arisaema triphyllum*) are perennial herbs widely distributed in temperate North America, from Nova Scotia, Ontario, and Minnesota south to Florida and Louisiana. Generally occupying moist and shaded areas, like the nutrient-rich slopes along streams, the plants produce three- to five-parted leaves from a rounded corm. The corms, when dried or boiled, served as an emergency foodstuff for Native Americans and early settlers, giving the plant the common name "Indian turnip." In addition, a salve for treating ringworm and snakebite was made by mixing the raw corm with lard. (WARNING: Raw corms contain large amounts of calcium oxalate crystals which, if ingested, cause intense pain and irritation of the throat's mucosal membranes.)

Jack-in-the-pulpit leaves and inflorescences emerge from the forest floor litter in late spring. An inflorescence consists of a slender stalk (*spadix*) surrounded by a green- or purple-striped sheath (*spathe*); the latter droops protectively over the top of the entire structure, like the canopy of an old-fashioned pulpit. Borne on the lower portion of the spadix are fifty or so highly reduced flowers—all male, all female, or (occasionally) both. Upon release, the light pink pollen collects in the bottom of the cup formed by the spathe; tiny insects called thrips (*Heterothrips arisaemae*) drench themselves in pollen, carrying it to a neighboring plant. The resulting red berries ripen by early autumn and are consumed by woodland creatures, such as box turtles, with the seeds disseminated in their droppings.

Although posing as simple herbs, jacks actually follow a complex sexual scenario—the Size-Advantage Model—similar to some advanced animals. According to this model, an individual will "choose" to be either male or female depending on the reproductive advantage to be gained; the amount of stored nutrients is a major determining factor in this choice. Large, healthy individuals, capable of sustaining the offspring, will be female; smaller and less healthy ones, required only to produce gametes, will be male; and marginal ones will "choose" not to reproduce at all, but concentrate instead on building up reserves for the next season.

In jacks, this all makes for a confusing (but predictable) situation known as sequential hermaphroditism. From one year to the next, an individual plant may change from asexual to male or female, from male to asexual or female, or from female to asexual or male. Size, rather than history or age, is the best predictor of sexual state, since a plant's size is a measure of its photosynthetic health—stored in the corm as starch—during the preceding year. Robust plants can "afford" to be female and support two leaves; males have only one leaf but are larger than asexual plants. Consequently, the females of a population usually develop on the most favorable sites, with males and asexuals on the harsher edges.

So the microenvironment, operating through the modification of endogenous hormone levels, must be considered as the ultimate determiner of gender in jacks. This environmental monitoring system ensures that the burden of fruit production is borne by the healthiest individuals, increasing the likelihood that genetic material will be transmitted to the next generation.

Jacks-in-the-pulpit, then, are "jacks" only when living less than ideal lives. As in other species, the most vigorous members of a population—and the ones bearing most of the reproductive costs—are the "jills."

Think in terms of the costs involved with reproduction. Find a tree "in fruit" and estimate the number of fruits produced. Multiply the number of fruits by the average weight of each one and calculate the total weight of fruits produced by that tree. What tiny percentage of those fruits will likely lead to successful reproduction?

Our painted bunting friend, having made it safely six hundred miles across the Gulf of Mexico, breathes his last. Photo by W. Mike Howell.

Birding Dauphin Island

*The Gulf Coast is an amazingly diverse and dynamic place,
especially during spring and fall migrations. A given day can show
birds falling out of the sky in huge numbers—and met by a similar
number of bird watchers. Some species, like the painted bunting,
have come to symbolize the beauty and wonder of these timeless
cycles.*

We migrated south that blustery April day, flying down
the highways in our minivans and SUVs, hoping to col-
lide with a more traditionally winged migration heading
north. Destination: Dauphin Island.

Five miles off the coast—and the first Alabama footfall
for the French in 1699—L'Isle Dauphin serves as a classic barrier island, its dunes and sandbars absorbing the
brunt of storms and hurricanes while protecting the fragile
coastline and its teeming estuaries. (Following this natural
cycle—and despite homeowners' complaints—big chunks
of the island often end up many miles elsewhere.) But to
migrating fowl, the isle creates a vital stopover spot—the
last chance to top off essential energy reserves before cast-

ing off on a six-hundred-mile, nonstop crossing of the Gulf of Mexico in the autumn, and the first chance to rest and recuperate from an equally exhausting flight in the spring.

Such migration defies human imagination. Birds—in some species, just a few ounces of feathered wonder—strike out on a timeless journey across a trackless oceanic expanse, wings beating rhythmically and constantly, beat after beat after beat propelling them (hopefully) ever landward. Fast species complete the journey after a grueling twenty-four hours; slower species take twice as long.

But why our accompanying human migration that spring day? In April 2004, an astounding 185 avian species had landed on Dauphin Island (earning it the title "America's Birdiest Coastal City"), and we hoped to spy a similar number. (Like other aspects of American culture, bird watching has become a competitive sport.) Above all, the mere glimpse of one species consumed us: the resplendent and much celebrated male painted bunting. Outrageously handsome even by ornithological standards, these elegant creatures sport purple heads and necks, yellow-green backs, and scarlet eye rings, rumps, and underparts—a veritable Mardi Gras of colors. (One of their Louisiana common names, *nonpareil,* tells it all.) Like other migrants, painted buntings tarry only briefly on the island, continuing on to interior thickets and hedgerows to raise their young and escort them back south in the autumn.

So we birded the most storied places, all part of the fifty-site Alabama Coastal Birding Trail, a mix of marshes, woodlots, and parks supporting migrant, vagrant, and resident species. (According to the official birding lexicon, migrants simply pass through on their way to summer or winter homes, while vagrants arrive totally unexpected—perhaps disoriented western species way off track. Residents, such as cardinals, neither come nor go.) The Shell Mound—an ancient midden pile with equally ancient live oaks—offered a haven for scarlet tanagers, blue grosbeaks, and indigo buntings. The Goat Trees, aptly festooned with Spanish moss, provided protective cover for shimmering gold prothonotary warblers. And the Airport Marsh harbored a multitude of ducks, coots, egrets, herons, and rails.

Our fellow birders constituted an equally motley flock—different stripes of human life, united by the overwhelming desire to view the most and rarest species. Easily identified, adults strutted forth in khaki shorts, hiking boots and knee-length socks, T-shirts supporting various conservation causes (with the tails flapping out), earth-colored vests with multiple pockets for field guides and notepads, super-sized binoculars, and digital cameras with humongous telephoto lenses. Conversations were whispered and crisp, with information freely shared. ("Red-eyed vireo, sapling to the left, about three o'clock.") Some boisterous types uttered witty birdicisms. ("I've been up all night, birding my candle at both ends.") Fledgling birders seemed quite scarce, while nestlings—proudly borne in camouflaged backpacks—were all named Robin.

But alas! Despite our collective efforts, we spotted no painted buntings. So we rested on the picnic tables of an abandoned Jiffy Mart, replenishing our own food reserves while reviewing the day's sightings. Then KA-THUNK! The sound reverberated across the concrete, ominously informing us of a much anticipated arrival. There on the weathered window sill lay a male painted bunting—purple, yellow, green, and scarlet—exhausted by his all-day-all-night quest and confused by the shiny plate glass. We picked him up and held him as, muscles warmly strained and neck mortally broken, he breathed his last. How ironic that this exquisite aviator survived seemingly endless sea before gaining landfall, only to thwack into a worthless human obstruction!

And while cradling our new friend, we offered this simple prayer: "Next year, may your relatives and loved ones, following the same ageless course, suffer happier fates. May they find safety and sustenance on the Alabama Coastal Birding Trail, resting adequately while restoring their much depleted energy. And may they move quickly on to inland summer homes, hale and hearty enough to find mates and raise many clutches of healthy chicks, who will in turn successfully negotiate the long and treacherous trek across the Gulf, to return with each succeeding spring. Amen."

Take a seat near a bird feeder on a perfect spring day, with binoculars and field guide close by. Note the species that you see and, for each one, whether it is a resident bird or spring migrant. (The field guide will have maps to show summer and winter distributions.) What percentage do you find in the two groups? Select one of the migrants and research its habitat and food needs. Where will it choose to breed and why?

_____

_____

_____

_____

_____

_____

_____

_____

_____

_____

_____

_____

_____

_____

_____

_____

_____

_____

_____

_____

_____

Teeth of prehistoric goblin sharks (*Cretoxyrhina*) from Shark's Tooth Creek.
Photo by L. J. Davenport.

chapter thirteen

Shark's Tooth Creek

Fossils tell their own type of stories—stories of ancient creatures living for eons under almost unimaginable conditions—and all told from remaining bones, teeth, and assorted fragments. Finding fossils is much like discovering buried treasure!

Most folks enjoy a good pirate yarn, filled with cold-hearted, steely-eyed swashbucklers, their pitiful and luckless "prey," plus the mysterious (and, perhaps, imaginary) buried treasure. But those folks lucky enough to stumble into Shark's Tooth Creek can dig up and cradle *real biological treasure* jettisoned by bloodthirsty "pirates" eighty million years ago.

The biological bullion of Shark's Tooth Creek was discovered quite by accident, in one of those serendipitous events that often propel science. The story goes that, on a particularly sweltering summer day, some paleontologists were rapidly approaching heat exhaustion at a nearby mosasaur (a huge marine reptile) fossil site. So, to cool off, they cautiously eased down the mucky banks of the creek into its cottonmouth-infested waters. (The number of snakes has risen exponentially with the retelling of this tale.) And, as their aching bodies gratefully sank into its chilly depths, the secret treasure of Shark's Tooth Creek seemingly nipped at their fingertips. Sharks' teeth! Thousands of *really old* sharks' teeth!

Now, compared to other creeks, Shark's Tooth Creek is not all that impressive—just a steep-banked, flat-bottomed ditch wandering lazily through the chalky clays of Greene County, Alabama. But, in those wanderings, it cuts through some of the premier fossil beds of the world. The rock base, known geologically as Selma Chalk, is formed from the calcium carbonate shells of tiny marine creatures called coccoliths, indicating that, during late Cretaceous times, shallow seas covered what we now know as the Black Belt, with Selma sitting at the top of a prehistoric Gulf of Mexico. (Coccoliths are highly varied and diagnostic of certain geological formations; similar deposits of similar age formed England's White Cliffs of Dover.) These warm, productive seas and coastlines provided ideal conditions for a rich community of algae, zooplankton, and larger invertebrates as well as a tremendous diversity of predators and superpredators, like mosasaurs and sharks.

While mosasaurs vanished long ago, sharks remain very much with us and are basically unchanged from ancient times. Ruthless marauders of the briny deep, these jagged-toothed, cold-eyed, conscience-deprived killers terrorized and feasted on small creatures that ventured too close. With this rapacious lifestyle, sharks constantly lost (and then replaced) their teeth, with those teeth accumulating in the bottom sediments. The sediments, after eons of compaction and consolidation, were lifted by the buckling of the Earth's surface. And even a small creek, like Shark's

Tooth Creek, cutting into those deposits, will concentrate teeth and other fossils in its bed.

All you have to do is pick the right spot, scoop up some sediment, and shake it through a wire screen. Oftentimes, two or three perfect teeth will appear. Because of their similarities to modern forms, the teeth of prehistoric sharks can be easily identified—makos, sand sharks, and lemon sharks are well represented. (A word of caution, though: the teeth may still be sharp!) You might even luck up and find a molar of *Ptychodus* as well as its lunch; these ten-foot prowlers, with grinding teeth an inch broad, dined on the colorfully (or disgustingly) named devil's toenail oyster.

So it's like a treasure hunt—just follow the clues and you'll be rewarded. Except that I disguised the name of Shark's Tooth Creek. (You won't find it on any map.) And the treasure is a geological one, with no *X* marking the spot. And the treasure has been buried for eighty million years. And the eye patch is optional.

Join a university- or museum-led fossil expedition. Take notes on all background information: geological formations, time periods, and major fossil finds. Then dig! Draw the fossils that you find and describe their shapes and colors. In which geological stratum were they found, and what is their age? Can you reconstruct or envision an entire plant or animal from fragments and footprints?

This *Lampsilis* inhabits the upper Cahaba River in Jefferson County, Alabama. Photo by W. Mike Howell.

Sex and the Single Freshwater Mussel

*The life cycles of freshwater mussels hold much fascination,
especially with their intimate ties to fish hosts. Add to that the
subtle variations of their beautiful shells, and you've got yourself
a lifelong hobby.*

*I broke a cardinal rule of nature writing by giving my mussel
an actual voice, but I think it helps to get the major points across.
Plus, it allowed me to be slightly outrageous.*

Put yourself in the place of a freshwater mussel. There you
sit, wedged into the gravel of an Alabama river, rhythmically extending and contracting your muscular, hatchet-
shaped foot to stabilize yourself in the rushing current.
With that foot, you clamber a few feet per day in search of
clean water, tiny critters to filter, and safety from terrestrial
predators, such as raccoons and muskrats. (You probably
won't move more than a few hundred yards during your
entire lifetime.) The sun shines brightly down on the

water's surface above you, but having no eyes (and no brain to process visual information), you remain in perpetual darkness.

Your simple body plan perfectly matches your simple needs. Incurrent and excurrent siphons protrude between the two valves composing your shell, bringing in and expelling water, respectively. Thin gill membranes remove both oxygen and food (mainly microorganisms and organic bits) from that water, propelling the minute food particles by ciliary action to the tube of your digestive tract. A sheet of glandular tissue, the mantle, surrounds these innards, secreting the glistening nacre (mother-of-pearl) that so beautifully lines your shell. Exquisite colors may decorate your abode—silvery white to deep purple, even orange. Depending on your species, your expected longevity can approach one hundred years.

"Okay, fine. I'm happy as a clam. But what about my sex life?" you impatiently ask. Well, a few of you enjoy hermaphroditism, producing both male and female parts with which to impregnate yourself. (Would such be considered a clam-destine affair?) However, most of you "choose" one gender or the other. If male, then your internal organs produce vast quantities of sperm, which are released to the water, then taken into a female's body through her incurrent aperture and carried to eggs waiting in her gills. Sections of these gills then serve as brood pouches or *marsupia,* producing parasitic larvae (*glochidia*) to be discharged into the water column. There they must immediately attach to the skin, gills, or fins of a fish, or die.

"Hmmmmm," you astutely ponder. "Aren't the odds pretty stacked against my precious, precocious glochidia? Won't they just be washed downstream or gobbled up as chub chow?" True, unless you're very clever. And clever clams turn the tables on their piscine predators, packing their embryos into explosive *conglutinates*—mucous-encased bundles resembling flatworms, leeches, or midges. A darter who dares disturb these provocative masses winds up with a face full of bambinos, which happily hitch a ride upstream.

Let's say you're a female *Lampsilis* or pocketbook clam, a rather glamorous but cautious creature clamoring for even better odds for her offspring. Why not modify your mantle to mimic a minnow—eyespot, tail, and all? Then use your muscular mussel contractions to wave it seductively in the water. What curious crappie could resist a close-up look (and accompanying cloud of clinging clamlets)?

"But how might I ensure the absolute biggest possible payoff?" Then you'd better create a superconglutinate—a transparent, six-foot-long gelatinous tube with all of your glochidia stacked at the distal end, tethered to your body (or attached to a stick or rock) and twirling invitingly in the current, appearing as a sick sucker. When a bigger fish bites this alluring lure, its false eyespot explodes, and the clamsters climb aboard.

Your "young-uns" remain embedded in the tissues of the host for one to six weeks, changing little in size but developing all adult organs and structures. (Don't worry about the fish; it's more embarrassed than debilitated.) Then these tiny editions, less than half a millimeter in diameter, drop off to begin their independent lives. And, reaching sexual maturity in three to nine years, they'll (hopefully) repeat their parents' cycle.

"So, will my clam clan be safe at this point?" Not really. Humans consider some of you to be quite tasty. Just witness the mussel shells near Muscle Shoals left over from aboriginal clambakes. In such prehistoric days, Alabama's rivers and streams harbored the most diverse assemblage of freshwater mollusks in North America. But nowadays, fully 69 percent of you Alabama bivalves are listed out as either extinct, endangered, or threatened. The biggest culprit—flooding from hydroelectric dams—turned dynamic, heterogeneous waterways into a series of silted, shallow lakes. (Loss of species diversity is "currently" not computed in our energy costs.) While some of you unique unionids hang on for years in these pitiful situations, your offspring have no hope for survival. Extirpation, then, becomes a prolonged event, lagging decades behind the actual responsible factor.

More subtle culprits reign. Streamside deforestation (with accompanying loss of cooling shade) leads to higher water temperatures, while strip mining and over-development of watersheds result in choking sediment loads. (Clamming up, by clamping down on your shell, is only briefly effective.) And some prolific live-bearing Asian mussels (*Cor-*

bicula), which need no host fish, will aggressively muscle in on your territory, leaving little room for you natives.

"Speaking of hosts. . . . What if, for similar subtle reasons, my own disappears?" Well, you'd be caught downstream without a paddle(fish).

Walk along a sand bar of a major river, picking up the whole shells and fragments that you find. How many species are represented? Make a sketch of one of those species, then use a field guide to identify it and investigate its life history. What are its host fishes? How are its glochidia spread?

A periodical cicada, fresh from its thirteen-year developmental "nap," clings to a sapling in Jefferson County, Alabama. Photo by W. Mike Howell.

chapter fifteen

Periodical Cicadas

This piece is based on my one (so far) personal encounter with periodical cicadas. In 1998, Dr. W. Mike Howell and I were conducting a fish survey in western Jefferson County, Alabama, when we were taken aback by an incredible, eerie sound—like metal dragging on concrete. It was easy to follow that sound to its source, a woodlot of small trees teeming with cicadas. It was also easy to see that the woodlot had been timbered and severely altered during the thirteen-year cicada life cycle. So what two successive generations of insects experience may be totally different things.

Every thirteen years—*precisely* every thirteen years—near the beginning of May, the woodlands of Alabama waken to an eerie, incessant din. The strange, numbing noise is quickly traced to thousands of two-inch-long, black-bodied, red-eyed creatures that literally arise overnight, crawling up from deep below the soil surface to perch and sing. And just as suddenly, toward the end of the same month, they completely disappear.

Often mistakenly called "locusts" after their voracious biblical cousins, periodical cicadas are among the world's longest-lived insects. They spend most of those lives tucked safely underground, supping on the juices kindly provided by tree roots and slowly developing through five instar stages. But every thirteen (or seventeen) years, in response to two-million-year-old molecular triggers, they rouse from seeming slumber, crawl slowly to the surface, and shed their final nymphal skin.

Upon reaching this brief state of adulthood, male cicadas cluster in enormous aggregations called choruses, vibrating a pair of abdominal membranes to produce a droning "song" whose decibel (or annoyance) level can reach 100. (A New England naturalist, in 1669, complained of "such a constant yelling noise as made all the woods ring of them, and ready to deaf the hearers.") Females, attracted to the chorus, are quickly mated; they then slit twigs with their blade-like ovipositors and deposit tiny, white, spindle-shaped eggs. By the end of June, the eggs hatch into first instar nymphs, raining to the ground and burrowing toward the tree roots. Their parents' bodies—finally silent—remain scattered and decomposing above.

This bizarre life cycle lasts either thirteen or seventeen years—but nothing in-between—with the longer cycles reserved for more northern climates. The nomenclature of periodical cicadas, as standardized in 1893, places them in thirty broods designated by Roman numerals. (Starting that year, numerals were assigned by the year of appearance.) But not all broods have been occupied, and some have become extinct. Fifteen are currently recognized—twelve of the seventeen-year cycles (I–X, XIII, and XIV) and three of the thirteen-year cycles (XIX, XXII, and XXIII). Brood XIX occurs in Alabama and adjoining states, appearing most recently in 1959, 1972, 1985, and 1998. (If you would like to pause, at this point, to reflect on your own life and the changes that took place between those years, *please do.*)

Why would such weird "now-you-see-'em-now-you-don't" strategies evolve? The generally accepted idea is that periodical cicadas benefit from their periodical nature through predator satiation. There's safety in huge numbers; and, no matter how greedily birds, squirrels, turtles, snakes, and spiders dine upon them, they just can't eat 'em all. And by emerging in perfect synchrony, and by staying site-specific, adults are assured of plenty of potential mates in the immediate neighborhood. (While mistakes are made, the occasional straggler who wanders too far, temporally or spatially, is quickly consumed.)

But why the magic numbers 13 and 17? This question, long puzzling to evolutionary biologists, was finally satisfactorily answered in 1997 (the Year of Brood III, if you follow the Periodical Cicada Zodiac). According to theorist Jin Yoshimura, climate cooling during ice ages led to the slowing of nymphal development—cooler soil temperatures stretched the original seven- to nine-year cycle (found in some modern cicadas) to twelve to fifteen years in the South and fourteen to eighteen years in the North. Although breeding groups developed, their timing patterns were not precisely fixed; insects from neighboring broods could meet and mate, producing asynchronous hybrids bound for extinction. Patterns based on non-prime numbers—12, 14, and 15 in one group and 14, 15, 16, and 18 in the other—came into phase and hybridized too often. But those based on the prime numbers 13 and 17 rarely hybridized—only once every 221 years—and remained genetically isolated, perfectly synchronized and out-reproducing all others. In this way, *timely* patterns of numbers became *timeless* ones!

And speaking of time. . . . The next ear-splitting emergence of Brood XIX is scheduled for 2011. What changes will they find, in your life and mine?

Investigate the periodical cicada cycle for your area to determine its next scheduled appearance. In what developmental stage are the animals presently found? What changes need to occur before emergence? Locate some places where cicadas have emerged in the past and where they will again. And be ready!

_____

_____

_____

_____

_____

_____

_____

_____

_____

_____

_____

_____

_____

_____

_____

_____

_____

_____

_____

_____

_____

An opossum ogles the author on his back porch in suburban Birmingham, Alabama. Photo by L. J. Davenport.

chapter sixteen

Opossums

This piece is, once again, based on a true-life adventure, bloody paper bag and all. It reminds me that, while many have gone extinct, some animals have actually done quite well—perhaps even improved their lots—since the advent of human civilization. But their success has absolutely nothing to do with evolutionary advancement.

Early one Sunday morning—much too early, in fact—I wakened to an ominous knock at the door. I opened it to find a tearful woman standing there clutching a wet and lumpy paper bag. She thrust it toward me, sobbing, "I ran over her and she still has babies!" (Ah, the joy of being the neighborhood biologist!) Sure enough, the bag contained the corpse of a freshly flattened female opossum—warm, bloody, and inert, except that her belly pouch wriggled oddly with life. I gingerly reached inside that pouch, wondering just what kind of disgusting mess I might find, then tugged at the offspring attached to her teats. One, two, three. . . . And finally ten emerged—warm, soft, and miraculously unscathed.

And that's how I learned, *firsthand,* about marsupial mommas and their mammae.

Such creatures are most associated with Australia, where kangaroos, koalas, and their cousins evolved in perfect isolation, radiating out from an original primordial marsupial into all possible niches and assuming all possible roles—grazers, climbers, and even predators (like the Tasmanian Devil of cartoon fame). But 'possums took a different path, arising in North America about seventy million years ago before "extincting," then returning (via South and Central America) during Pleistocene times. These "living fossils" remain pretty much unchanged, among the Earth's oldest surviving mammals.

Primitively clumsy and slow, with conical heads and pointed pink noses, sparse gray fur and naked ears, mature opossums reach the size of house cats (about ten pounds). Nocturnal and arboreal, they scramble up trees and hang from limbs thanks to long prehensile tails and opposable inside toes (like thumbs) on their hind feet. Truly omnivorous, they eat all that they find—birds and their eggs, worms and snakes, frogs and toads, insects, nuts and fruits (showing a particular penchant for persimmons), vegetables, garbage, cat food on my back porch—the 'possumbilities are endless. They even happily consume carrion, wandering onto busy highways to get it. (So in a perverse Cycle of Life, roadkill begets roadkill begets roadkill, ad infinitum.) Perhaps, instead, they should order carrion carryout. . . .

Non-motorized humans also obliterate opossums. Long considered as "varmints," due to their depredation on hen houses and garden plots, 'possums are hunted for both sport and food. (When cleaning the carcass, be sure to remove the rather pungent sacs on the small of the back and under the forelegs.) Baked 'Possum with Red Peppers and Sweet Potatoes makes a particularly savory dish—or so I've been told.

The name "opossum" entered our language in 1608 thanks to Captain John Smith, who anglicized an Algonquian word (*apasum*) meaning "white animal." So the initial *O,* which we frequently drop, is part of the o-riginal spelling, leading to o-ccasional jokes. ("What's Irish and keeps her babies in a pouch? An O'Possum.") And O'tongue-twisters. ("Obese opossums often ooze offensive odors.") But the *O* blocks the alliterative perfection of PLAYING 'POSSUM! Early settlers marveled at the creature's propensity, when cornered or caught, to suddenly go limp, fall down, and roll over, with tongue hanging out, drooling profusely. 'Possum pursuers quickly lose interest in an apparently dead animal (*I mean, what's the fun in that?*) and leave it alone. This stress-induced catatonia, similar to fainting, lasts up to four hours, until the 'possum senses its safety, blinks open an eye, smiles to itself, and ambles away.

'Possum reproduction follows the marsupial party line. Males mature at eight months, and females even earlier. Once the two get together, gestation lasts less than fifteen days. The "embryos" emerge about the size of a navy bean, crawling blindly, hand-over-hand, from the womb to the pouch (marsupium), following a moist path licked in her fur by Mama. Thirteen nipples await—a circle of twelve with one in the middle—and the youngsters instinctively attach, swallowing the life-giving thread deep down their throats. There they remain, contentedly suckling, for two months, wriggling and crawling about as though attached to a tether, then cautiously letting go, peering out the pouch's door as their mother wanders about the countryside. The final month of childhood is spent riding on the back of this maternal mobile home, clinging tightly with their tails.

And by the way: Those ten orphaned opossums, about six weeks old, were whisked to the wildlife rescue folks at Oak Mountain State Park, who rounded up adoptive parents equipped for 'round-the-clock feeding. And they all made it just fine, released to the wild when ready. Hopefully, each lived a good, long life (three to five years); had many children (two to three litters per year); successfully avoided owls, bobcats, foxes, hungry humans, and unripe persimmons; and—most importantly—shied away from asphalt.

Choose another animal that has thrived—even expanded its range—
over the past fifty years. (Some suggestions are coyotes, raccoons, and
armadillos.) What makes this animal so much more adaptable than its
brethren? What special characteristics allow its continued existence near
(or even with) humans and their habitations?

_____

_____

_____

_____

_____

_____

_____

_____

_____

_____

_____

_____

_____

_____

_____

_____

_____

_____

A brown pelican floats near a Gulf Coast pier. Photo by W. Mike Howell.

Brown Pelicans

Brown pelicans make a happy story, pulling back from the brink of extinction. They also provide a lesson in limericks.

Poetry, even in its crudest form, sometimes captures biological truth. Dixon Lanier Merritt's 1910 limerick, for example, teaches us that a pelican can hold more fish in his beak than his "belican," and we pause to wonder, with the author, "how the helican."

And we *should* wonder at these comical yet majestic creatures. Pelicans inhabit the edges of oceans and lakes, with seven species worldwide. One species, the brown pelican, occupies the Atlantic, Gulf, and California coasts of the United States, extending its range south into the tropics. Adults weigh five to eight pounds, ably supported by a six- or seven-foot wingspan; eighteen-inch bills house distensible fishing pouches capable of encompassing three gallons of water. (That's three milk jugs' worth!) Juveniles wear nondescript mottled brown plumage, while adults display white heads tinged with yellow, the white extending down the neck in a narrow band; the rest of the neck is a handsome dark chestnut.

Gregarious birds, congregating throughout the year in flocks of both sexes and all ages, brown pelicans feed exclusively in shallow estuarine waters. Menhaden, a bony sardine or "rough" fish of little human use, forms their main fare. (Notorious panhandlers, tame pelicans supplement their diets by begging handouts at fishing piers and seafood shops.) Sand spits and offshore bars serve as day-time "loafing" and nighttime roosting areas.

On foot, with their rotund bodies, comic-character bills, and enormous webbed feet, brown pelicans appear as clumsy as any circus clown. But true grace awaits in the air. With slowly measured wing beats, heads curled back heron-like, barely skimming the tops of waves or sailing along the valleys in-between, they solemnly and superbly glide—steady, dignified, and confident.

Thanks to keen eyesight, brown pelicans can spot a single fish from a height of sixty to seventy feet. Wings folded, turning in a half roll, they suddenly plummet bill-first into the water, with pouches fully open, disappearing from sight in the resulting splash. (Always entering downwind, pelicans somersault underwater to emerge upwind for better liftoff.) Springing quickly to the surface and shaking moisture from their plumage, they point their bills downward to drain out excess water, then quickly upward to swallow their catch.

Breeding in congested colonies on lonely shores and isolated islands or "mud lumps," pelican pairs build crude nests on the ground or in mangroves. Newly hatched chicks—blind and featherless—are completely dependent on their parents to shade them with their wings and feed their growing bodies. The parents take turns at fishing, returning to the nest to regurgitate their catch back into the pouch. The nestlings, sometimes two at a time, eagerly crowd this parental feeding trough, thrusting downy heads deep into its innermost recesses where, with much squabbling and squawking, they savor a hearty, partially digested meal.

For brown pelicans, life remained unchanged until the late 1950s, when—like birds out fishing—their numbers suddenly plummeted. Between 1957 and 1961, along the Gulf Coast from western Florida to southern Texas, the species virtually disappeared. (Alabama's population dropped from eighteen hundred birds to a mere sixty; Louisiana, the so-called Pelican State, had none.) Little sleuthing was needed to discover the culprits behind this catastrophe—endrin and DDT. In a classic case of biomagnification, these pesticides became concentrated more and more up the food chain; pelicans, at the top, took in lethal doses with each mouthful of menhaden. Endrin worked its deviltry directly, poisoning or killing all age classes. The more insidious DDT, through its breakdown product DDE, interfered with calcium deposition during egg formation; pelicans that still managed to produce these paper-shelled products ended up crushing their offspring-to-be during incubation.

Biologists (and politicians) quickly rallied around these befuddled birds, listing them as an endangered species in 1970—three whole years before passage of the Endangered Species Act! The newly established Environmental Protection Agency flexed its muscles by banning endrin and DDT for all time. With acknowledged irony, wildlife workers successfully recolonized nesting sites in the Pelican State *using Florida birds*. And the birds themselves helped their cause. In 1983, several pairs of pelicans nested on Gaillard Island, a spoil bar in Mobile Bay; this colony—the first known nesting site *ever* in Alabama—soon developed into one of the largest brown pelican colonies along the Gulf. All of these efforts, human and avian, combined and culminated in 1985, when brown pelican populations in Alabama, Georgia, Florida, and the Atlantic Coast were happily removed from federal endangered status.

So, in homage to this homely yet magnificent (and resilient) bird, I offer the following updated version of Merritt's classic:

A wonderful bird, the brown pelican,

Whose beak can hold more than his belican.

Once barely surviving,

He's now healthy and thriving.

And with no DDT in his shell, he can.

Research the case of the bald eagle, a species similarly pulled back from the brink of extinction. What factors contributed to the bald eagle's plight? What changes in human activities led to its recovery? How similar or parallel are/were the situations of the bald eagle and brown pelican? Just for fun, condense the "bald eagle story" into a limerick!

Tiny yellow-green flowers reveal the bromeliad affinity of Spanish moss. Photo by L. J. Davenport.

chapter eighteen

Neither Spanish nor Moss

The title of this chapter was stuck in my brain for many years before I finally wrote it. It's about the misnaming so often found in nature study, and how you have to sometimes wade through wads of misinformation to get to the truth. Or maybe it's just about words that rhyme with festoon!

Festoon—a funny-sounding word referring to festival decorations and closely associated with Spanish moss, whose own name makes little sense. . . .

One of the world's most widespread and readily recognized plants, Spanish moss grows naturally from the southeastern United States to Argentina. (This five-thousand-mile latitudinal extent constitutes a modern-day botanical record!) Wherever found, its characteristic thread-like masses festoon (*There's that word!*) host branches, cascading toward the ground in long silvery streamers.

A fellow bromeliad, Spanish moss shares many characteristics with pineapples, including a special drought-resisting type of photosynthesis. Tiny yellow-green flowers produce inch-long capsules; the seeds inside bear hairy sails and float on the slightest breeze, lodging in the cracks of rough-barked trees, like live oaks. While flowers, fruits, and seeds contribute some, most propagation involves plant fragments carried from tree to tree by birds and winds. Spanish moss's unprecedented range through Mesoamerica, in fact, may be due to hurricanes.

So far, then, it's neither from Spain nor a moss; and, contrary to popular belief, neither is it a parasite. Spanish moss gains no nutriment from its host, as a parasite would. Instead, a dense indumentum of absorbing hairs traps the water and ions necessary for life. The host tree suffers no damage, although extra thick growth may block its sunlight. Humid air trapped within these masses ameliorates the harsh conditions prevailing outside, allowing Spanish moss to shelter a broad assemblage of creatures, including insects, snakes, and bats (which often fall prey to the snakes). Warblers use Spanish moss for nest building, and one spider, *Pelegrina tillandsiae,* makes it its sole habitat. (Although tempting, campers should refrain from using Spanish moss for bedding—chiggers like it, too!)

Our un-Spanish/not-moss/non-parasite also sports a confusing scientific name, *Tillandsia usneoides.* While the first part simply commemorates the seventeenth-century Swedish botanist Elias Tillands, the second part demands more study: *Usnea* is a lichen notorious for festooning tree limbs, and the suffix *-oid* means "resembling" or "kinda like but not quite," as in typhoid, celluloid, humanoid, or (*Dare I mention?*) hemorrhoid. So Spanish moss is kinda like a lichen, but—as in other things—it's not.

Together with live oaks, magnolias, and muscadines, Spanish moss symbolizes life in the American South; for journalist James J. Kilpat-rick, it "wraps the hard limbs of hard times in a fringed shawl." British naturalist Philip Henry Gosse, teaching in Alabama's Black Belt during the 1830s, also chose a sartorial theme, describing dead trees "stretching their gaunt white arms, clothed with long ragged festoons of Spanish moss." Native Americans called it "tree hair," which French explorers turned into "barbe espagnole" (Spanish beard) to insult their bitter rivals in the New World. According to legend—no doubt composed by the French—a particularly ugly, crude, and cruel Spaniard named Gorez Goz tried to purchase a native girl, chasing her up into a tree; on follow-ing, his mangy beard got hopelessly entangled. There he died, the tree forever festooned by the voluminous beard of a doubloon-laden Spanish buffoon. (Was he wearing pantaloons?)

In the story above, the native girl just might have worn a dress woven from Spanish moss, while the explorers and later colonists used a mix-ture of moss and mud to caulk their cabins. Dried moss served ably as kindling, while long strands repaired fishermen's nets. But other than garden mulch and packing material, most modern uses require modified rather than raw product, especially for stuffing upholstery. Gathered moss is soaked in water, then piled in heaps or buried in pits for up to six months; the outer "bark" rots away, leaving a black, resilient, wiry filament which is dried, ginned, and baled. Commercial shipments from the South began shortly after the Civil War, with over ten thousand tons cured in 1939—just before synthetic fibers bottomed out the market.

Folks also use Spanish moss for medicine and decorations. Colds, coughs, fevers, hernias, measles, mouth sores, dandruff, diabetes, arthritis, and ailments of the lung, liver, kidney, and heart are treated with various concoctions. Astute viewers will notice Spanish moss adorning the sets of Tarzan movies, belying the films' purported African setting. And, returning to the origin of "festoon," South Americans use it as Christmas ornaments and to cradle the Christ child in nativity scenes.

To help us appreciate this widespread, useful, misunderstood, and misnamed plant, I end with the following bit of doggerel:

Spanish moss is many things, a wonder to behold,

A pendant form of pineapple, treatment for a cold.

Stuffing for upholstery, cabin caulk and dental floss,

But these two things it's simply not—Spanish or moss.

Choose a plant that is indigenous to your area and investigate the origin of its common name. (Suggestions include French mulberry, Osage orange, and New Jersey tea.) Does that name contribute to our understanding of the plant, or does it detract from that understanding instead? If you had no previous knowledge of the plant's name, could you come up with a more reasonable one that would better reflect its biology?

_____

_____

_____

_____

_____

_____

_____

_____

_____

_____

_____

_____

_____

_____

_____

_____

_____

_____

An ocellated flounder, mouth agape, peers from its tank at the Dauphin Island Sea Lab Estuarium, Mobile County, Alabama. Photo by W. Mike Howell.

Flounders and Other Flatfishes

This chapter is about the amazing transformation in the development of flatfishes. And it's also about seafood.

"Step right up! Right this way! See the fish with eyes on the TOP of its head! This is a totally amazing freak of nature!"

Not even a carnival sideshow could fabricate the amazing freaks of nature known as flatfishes. Gliding through the ocean like finned flying carpets, they settle to the bottom, flip sand on their "backs," and completely vanish—except for bulging eyes. There they lie, biding their time as the watery world scoots and scuttles above. Then, in a twinkling of those eyes, they burst forth in jet-propelled frenzy to gobble up tasty prey. We humans, in turn, pursue flatfishes with seines, hooks, and gigs, seeking most flavorsome halibut, flounder, and sole.

Unlike other flattened fishes, such as stingrays, flatfishes start off life perfectly "normal," with all parts oriented up and down, left and right. But soon after birth, a flatfish larva tilts on its side; simultaneously, its skull grotesquely warps, as the downward eye—plus its socket and optic nerve—*picks up and moves* 120 degrees around the top of the head. (A twist of fate? What kind of twisted thoughts would this fish be thinking?) After a few days, both eyes peer quizzically from the "top"; the mouth, likewise distorted in the process, wears an appropriately crooked expression. While their blind sides remain pale, eyed sides simulate the colors and patterns of hiding places.

But this is not, by any means, a one-sided story. The combination of two starting points and a single roaming eye results in left-eyed and right-eyed taxonomic groups. In general, cold waters produce right-eyed flatfishes, like the North Atlantic halibuts and soles, while warmer waters yield left-eyed creatures, like flounders and tonguefishes. (While "filet of sole" originally referred specifically to the European food fish, in the southeastern United States the term applies to any edible form.) About three dozen flatfish species are known from the Gulf Coast, with flounders showing the greatest diversity, especially in size and color pattern: southern, Gulf, broad, three-eyed (due to the presence of a prominent spot, not an extra eye), fringed, and ocellated.

Of these, the southern flounder is the most important food and sport fish; therefore, much is known about its life cycle and habitat needs. During late autumn, southern flounders leave their protected coastal waters for spawning in the open Gulf, generally at depths of fifty to one hundred feet. For the resultant larvae, eye migration is complete by the time they reach one-half inch long. Young fish enter bays during early spring, seeking shallow, grassy areas near Gulf passes, then ease up coastal rivers and bayous. Small flounders grow rapidly, reaching twelve inches by the end of the first year; males seldom exceed this length, while females double it. (In halibuts, this sexual dimorphism is most extreme—females may reach eight feet long and weigh hundreds of pounds, with males one-tenth as large.) Most southern flounders caught (and happily consumed) are females twelve to sixteen inches long, weighing one to two pounds.

Juveniles feed mainly on crustaceans, but small fish become more important in the adult diet. An adult flounder enters shallow water at night and wriggles into the sand and mud, hoping that potential prey will wander close. Then, by forcefully shooting water out its downside gill, it explodes from the bottom, using razor-sharp teeth to dispatch its victim, then settles gently back down again. Fishes like flounders, that strike fast but have no staying power, possess flanks of mild-flavored and flaky white muscle that lend themselves most admirably to the culinary arts. (The flight muscles of "flightless" birds, like chickens, are similar.) Broiled, baked, or fried, perhaps stuffed with crabmeat or scallop dressing. . . . Mmmmmmmmm good!

Because of their lightning-quick reflexes, southern flounders make storied game fish. But an angler who doesn't have his rig (and attached arm and shoulder) jerked overboard during the first vicious strike usually has little trouble landing his soon-exhausted, shifty-eyed catch. (Experienced fishermen use live bait—shrimp and killifish—to ensure success.) The more sedate gigging or "floundering" is best done by lantern light around jetties and oyster reefs, especially on calm, dark nights during late autumn. Hefting a three-pronged, long-handled spear, the flounderer wades the periphery of salt marshes or poles tidal creeks in a shallow-draft boat. He cautiously approaches the camouflaged outline of a buried flounder, then squarely "plants" his gig through its top side. (The lantern also illuminates a multitude of fishes, crabs, and other marine life—a spectacular nighttime natural history treat!) Novices should be careful, though, to avoid gigging stingrays.

But like the carnival barker at the beginning of this story, I'm running out of time (and space) to sell you on the wonders of flatfishes. (Or perhaps my attempt is *floundering*. . . .) So I leave you, patient reader, with two awful but instructive puns: Would a flounder have no sole? And does this weird eye migration happen just for the halibut?

What's your favorite seafood? Investigate the fish that provides it, describing its life cycle in detail. Is it entirely deep-sea or also estuarine? When (and why) during its life cycle is your fish best harvested? Are there any stresses or limits on its future consumption?

A liverleaf enjoys early spring weather in Jefferson County, Alabama.
Photo by W. Mike Howell.

Liverleaf and the Doctrine of Signatures

Plants and medicine have been intimately linked throughout human history. This chapter explores the philosophical basis for such, using a common spring wildflower.

Some belief systems embrace the universe, uniting all living things with the very heavens above. Liverleaf, a common North American and European wildflower, embodies one such "truth."

Compared to its more spectacular springtime colleagues, liverleaf (or hepatica) appears none too special—just a few white, pink, or lavender blooms popping up through mottled, coppery leaves left over from the previous year. But those leaves possess "powers" unparalleled in the botanical world. The ancient Greeks used them to treat known liver dysfunctions, plus two other suspected symptoms of such disease: cowardice (*"Yep! I'm callin' you yella, Hepatocles!"*) and freckles. For Native Americans, liverleaf tea aided bilious ailments and generally poor digestion. In American folk tradition, the tea was used for fevers, jaundice, and coughs. And a brief "liver tonic" boom resulted in the consumption of 450,000 pounds of dried hepatica leaves during 1883 alone!

But herbalist/apothecary Nicholas Culpeper (1616–1654), author of *The English Physitian, Or an Astrologo-Physical Discourse of the Vulgar Herbs of this Nation,* reigns as liverleaf's all-time greatest fan. To wit: "It is a singular good Herb for all the diseases of the Liver, both to cool and clense it, and helpeth the Inflamations in any part, and the yellow Jaundice likewise: Being bruised and boyled in small Beer and drunke, it cooleth the heat of the Liver and Kidneys. . . . It is a singular remedy to stay the spreading of Tetters, Ringworms, and other fretting and running Sores & Scabs, and is an excellent remedy for such whose Livers are corrupted by surfets wch causeth their bodies to break out, for it fortifies the Liver exceedingly and make it impregnable."

Culpeper's conviction that liverleaf cools, cleanses, restores, rejuvenates, and fortifies this organ—perhaps even *cures* every hepatic complaint ever recorded by humanity—was based solely on its leaves being three-lobed and reddish-brown, just like the human liver. Plants that LOOK like certain parts must be GOOD for those parts—the Doctrine of Signatures. Simply put: (1) God graciously created plants with divine powers that will ease our many miseries, but since (2) we humans are too dim-witted to figure these things out for ourselves, (3) God kindly provided clues—signs or signatures—to show us the way. If we just follow those signs, good health results. (Most intriguingly, this comforting belief crosses cultural and religious lines—Native American, Occidental, Oriental, Ayurvedic. Different gods, but the same "truth.")

So we use walnuts, which look like brains, for epilepsy and mental illness; bloodroot, with its red sap, figures in circulatory problems; the fused leaves of boneset heal broken bones. Long-lived plants help us live longer, while those with thorns ease our pains. And ginseng—its bifurcated root resembling an old man's legs—*must* stimulate the aging parts in-between.

In our Western world, the Doctrine is usually attributed to the Swiss alchemist/physician/philosopher/gadfly Paracelsus (1493–1541), contemporary of Nicolaus Copernicus, Leonardo da Vinci, Martin Luther, and a host of equally iconoclastic figures. Born Phillippus Aureolus Theophrastus Bombastus von Hohenheim—perhaps "Bombast" for short—Paracelsus rode the Renaissance wave of humanistic rediscovery. His two favorite ancient authorities were Hippocrates of Cos, whose medicines mimicked the diseases being treated, and Aulus Cornelius Celsus, with his elegant medical Latin. Egotistically adopting a moniker meaning "above Celsus," Paracelsus crucified the reigning medical establishment, ceremoniously burning the standard texts and methodically reducing compounds to their bare essences. "Like cures like!" he cried, even using minuscule amounts of lead and mercury to treat such poisonings!

German spiritualist Jakob Böhme (1575–1624) picked up Paracelsus's notions and shot them to new heights. A shoemaker by trade, he experienced "a profound mystical vision" that revealed to him the *true* relationship between humans and God. (An insight while pondering an insole?) According to Böhme, the macrocosm (heavens) reflects the microcosm (humans); to study one is to know the other. *As above, so below;*

as without, so within—this is the centerpiece of Renaissance cosmology.

Culpeper took all this and fused it—alchemy, spiritualism, and astrology, adding his own pharmaceutical slant. He explaineth thusly:

> I knew well enough the whol world and every thing in it was formed
> of a Composition of contrary Elements, and such harmony as must
> need shew the wisdom and Power of a great God. . . . I knew as well
> this Creation though thus composed of contraries was one united
> body, and man an Epitome of it, I knew those various affections in
> man in respect of Sickness and Health were caused Naturally . . . by
> the various operations of the Macrocosm; and I could not be igno-
> rant, that as the Cause is, so must the Cure be, and therefore he that
> would know the Reason of the operation of Herbs must looks up as
> high as the Stars.

According to this Doctrine, a *physitian* carefully matches diseases and their cures with the Zodiac, thereby balancing the universe within each patient. And herbs, like liverleaf, securely connect us to the Creator and the cosmos. After all, there's just a one-letter difference between *planets* and *plants*.

Choose a common wildflower from your own region and research the
history of its medicinal uses. (Suggestions include boneset and bloodroot.) How does its name "fit" the Doctrine of Signatures, indicating its
uses? Do current scientific studies support its effectiveness in such treatment?

_____

_____

_____

_____

_____

_____

_____

_____

_____

_____

_____

_____

_____

_____

_____

_____

_____

_____

A green tree frog sleeps peacefully on a sumac branch in Lawrence County, Arkansas. Photo by L. J. Davenport.

Green Tree Frogs

Dr. W. Mike Howell and I were on a field trip to the Delta region
of Arkansas when we spied some tree frogs sleeping the day away in
a sumac patch. And seeing those frogs made me wonder about their
calls and communication, and just what the nighttime might bring.

As spring hurries toward summer and evening temperatures rise, so does the decibel level at southern ponds and
puddles. Every night, in a ceremony that pre-dates the
dinosaurs, male frogs and toads congregate at the water's
edge to grunt, moan, whistle, or trill their way (they hope)
into a female's heart. Often, a key voice in this primal chorus is the "quonk" of the green tree frog.

Green tree frogs (*Hyla cinerea*) have been described by
naturalists as "graceful," "elegant," "aristocratic-looking,"
and "the most beautiful tree frog of North America."
Considered large by tree frog standards, they may reach
two-and-a-half inches from snout to vent. Their bodies
are brilliant green, although the color may change to dull
green or slate gray when they are hidden or inactive; the
specific epithet (*cinerea*) means "ash colored" and is based

on this change. No matter the body color, each side of the frog is marked by a straight yellow or white stripe, with a similar stripe appearing on each leg. The tips of the digits are expanded into adhesive disks, allowing the creatures to climb vines, shrubs, and small trees.

Green tree frogs are native to the coastal plain of the eastern United States, from Maryland to peninsular Florida, across to Texas, and up the Mississippi Valley to southern Illinois. In Alabama, they are typically encountered from the central zone to the Gulf Coast, with a few isolated populations in the Appalachian Plateau.

Lakes, ponds, swamps, and streams are prime habitats for green tree frogs, especially those places with abundant emergent vegetation. During daylight hours, they may be found sitting quietly or sleeping on leaves or stems, with their legs folded beneath them; they are also known to hide under loose bark and in tufts of Spanish moss. A "sit-and-wait" foraging strategy is utilized, with frogs leaping to spear flies, beetles, and spiders. Some individuals have learned to frequent windows at night, where they seek insects attracted by the lights. Green tree frogs, in turn, become prey for large aquatic birds, snakes, raccoons, and fishes.

In the Southeast, male green tree frogs sing from April to August, perched on cattails or reeds one to three feet above the water. Their courtship call—a ringing, nasal "quonk" or "quank"—is repeated in a measured cadence, thirty to sixty times per minute. Females, attracted by the calls, move toward the males and are soon mounted; from three to five hundred brown-and-cream fertilized eggs are released per pair. After a few days, the eggs develop into tadpoles, which two months later metamorphose into mature forms.

While the above follows the basic amphibian reproductive scenario, green tree frogs are known to throw in a major twist. Males engage in "sexual parasitism," an arrangement in which calling males are accompanied by one or two silent satellite males. When a female moves toward a calling male, a satellite will intercept her and mate with her. (Parasitism is involved because the calling male risks his life by attracting predators, yet gets nothing for his efforts.) In a further wrinkle, males may change from calling to satellite status and back again, on alternating nights or even on the same night.

In terms of reproductive success, green tree frogs appear to be doing far better than many of their cousins. Biologists have noted a dramatic recent decline in amphibian numbers, with both local populations and entire species seeming to disappear overnight. Many explanations have been proposed: destruction of wetlands and rainforests, pesticide pollution, heavy metal contamination, fungal infection, global warming, acid rain, ozone depletion, and increased ultraviolet radiation—even the harvesting of frog legs for food! While the scientific jury is still out as to why the decline has occurred, the decline remains very real. Amphibians, in fact, may be replacing birds as monitors of environmental health, forming an early warning system signaling both imbalances and advanced degradation.

But if noise levels at certain wet spots can be trusted, green tree frogs are doing just fine. As long as there are males going "quonk" in the night—plus some satellites to do much of the mating—the species should survive.

Pick a species of tree frog that is native to your area and research its life cycle and behavior. Obtain a tape recording of its mating call and describe it in such a way that you can readily imitate it—or at least recognize it. Visit a local pond during springtime and listen carefully for "your" frog. Do you dare to join in the chorus?

Cedar apple rust galls adorn a cedar tree near Russellville, Franklin County, Alabama. Photo by L. J. Davenport.

chapter twenty-two

Cedar Apple Rust

Fungal life cycles are particularly fascinating in that they sometimes require a second host. Such is the case with the aptly named cedar apple rust.

One drizzly spring day, while wriggling through a thicket in Franklin County, Alabama, I came face-to-face with a most horrible sight: an entire cedar tree "eat up" with orange oozes, dripping dew like a thousand noisome noses. But—intrepid biologist that I am—I felt no fear, for I only faced the teliohorns of cedar apple rust (CAR).

Beginning with Cedar: Like other rust fungi, CAR demands two hosts to complete its life cycle—cedar trees and apples—and in very precise and particular manners. First, cedar trees support its gruesome galls—roundish, reddish-brown, golf-ball-sized, and similarly dimpled. With warm spring rains, swollen, horn-like fingers descend from those dimples, dangling increasingly downward and outward with each successive shower—bright yellow-orange tentacles two inches long. During six to eight swelling/drying cycles, a single gall discharges two billion spores, its protuberances finally withering into exhausted, wrinkled threads.

On to the Apples: During the exact same time, nearby apple leaf buds pop open, hopefully (from CAR's selfish viewpoint) providing perfect substrate for spore lodgment and growth. (The fungus prefers several hours of light rain, at 50 to 75 degrees Fahrenheit, falling on tender, succulent, four- to eight-day-old leaves or young fruits.) While the wind-blown spores may wander five miles from the "mother" cedar, most apple infections occur within a few hundred feet. There they morph into copper-colored pustules, each with a characteristic chlorotic halo—as many as three hundred spots on a given leaf. Within these putrescent pools, black *pycniospores* fuse to form hyphae growing toward the opposite leaf surface.

An Etymological Aside: The word *heteroecious,* combining Greek words for "different" and "house," perfectly describes a rust's life cycle, alternating between two unrelated hosts. (Smuts, the other main type of plant infection, demonstrate an *autoecious* or self-housed cycle—e.g., corn smut grows only on corn.) The suffix also helps form *ecology,* the study of the house where we live, and *economics,* household management.

Apples, Part Two: Our fungal strand, now massed on the lower leaf surface, erects quarter-inch-long structures (*aecia*) directly below the original coppery spots. During the warm, moist weather of late summer—with a temperature of 75 degrees being optimal—these elongated sacs split into narrow strips and curl backward, releasing chains of airy brown *aeciospores,* which drift into the cracks and crevices of neighboring cedar twigs. It is in this form that CAR overwinters before growing (through the subsequent year) into the next generation of dimpled galls.

A Brief Review: The cycle is teliospores (on cedars), pycniospores (on apples), aeciospores (on apples), and back again. These are all manifestations of the same creature, but inhabit different (and specific) parts of different (and specific) plants at different (and specific) times.

Breaking the Cycle: An orchard full of CAR proves quite "galling" to apple growers, defoliating and weakening trees and blemishing their fruits, rendering the latter unmarketable. So why not nip it in the bud—quite literally—by disrupting the heteroecious cycle? This calls for total eradication—every cedar within a spore's throw of apple orchards, plus wild apples and (likewise susceptible) native hawthorns. *All must go!* But what about the pretty pink crabapple in your mother's yard, or your own perfectly placed foundation planting of ornamental junipers? Just where does the Garden Gestapo stop? It is the very teliohorns of a dilemma. . . .

While walking in the woods, though, there is no need to be frightened by an occasional gelatinous tendril. Unless, of course, you share a lot of apple DNA and your leaves are only four to eight days old.

Research the life cycle of wheat rust, another heteroecious fungus. What are its hosts, and what is the interplay of spores between those hosts? Use the format of the above chapter to map out the details of the wheat rust life cycle. What efforts are made to stop the spread of this fungus, especially in wheat-growing regions?

_____

_____

_____

_____

_____

_____

_____

_____

_____

_____

_____

_____

_____

_____

_____

_____

_____

_____

_____

_____

_____

An adult gopher tortoise captured on a feeding foray in Henry County, Alabama. Photo by L. J. Davenport.

chapter twenty-three

Gopher Tortoises

The longleaf pine ecosystem is one of the most endangered of the southeastern United States. This chapter is about one key inhabitant of that ecosystem and the vital roles that it plays.

Sometimes a hole is more than just an empty space. For a gopher tortoise, its hole or burrow is home, sanctuary, courting ground, and incubator—its very life.

"Gophers" are native to droughty, deep sand ridges from South Carolina through Florida and west to Louisiana. Open, savanna-like expanses of longleaf pine and scrub oak form the tortoise's preferred habitat, but they will readily colonize less pristine places like fencerows, pastures, and field edges. Like all good grazers, they contentedly consume the grasses, legumes, and fleshy fruits within their short reach, dispersing (and fertilizing) the seeds with their droppings.

Perfectly adapted for digging into sandy soil, gophers possess massive front feet which end in five flattened toenails. By scraping ahead and pulling quickly back, they send the sand flying in rhythmic spurts. Their rear legs are elephantine—round and rubbery, ending in flat, claw-equipped pads—providing both stability and strength. With this powerful earth-moving equipment, gophers can tunnel forty feet, the holes slanting twelve feet down, with a convenient turn-around at the end.

Often described in human terms as a mixture of gentleness, uncon-cern, stubbornness, and intelligence, gophers spend the majority of their lives relaxing in their burrows, perhaps in quiet contemplation. Here they enjoy protection from fire, predators (such as feral dogs), desicca-tion, and temperature extremes. Other species find the burrows inviting, with one report noting an assortment of sixty vertebrate species and three hundred invertebrates sharing these accommodations. Common bedfellows are gopher frogs, eastern indigo snakes, and eastern dia-mondback rattlers.

While gophers spend most of the winter months holed up in bur-rows, their activity increases with warm spring temperatures. By sum-mer, their mid-morning and mid-afternoon feeding forays prove taxing enough to prompt a much-needed siesta in-between. Foraging activities start and end at the burrow, the gophers trudging along a well-worn cir-cular or elliptical path, never venturing too far afield; in fact, 95 percent of all feeding takes place within thirty yards of "home."

This home is not permanent, however, and adults pick up and move close together in colonies during the breeding season, April to June. A dominance hierarchy soon develops, with the largest males (ten to fif-teen inches long) and breeding females monopolizing the colony's cen-ter. Courtship consists of a male "calling" at a female's burrow entrance, bobbing his head to better waft his scent, then gently biting her when she emerges. (Because the two sexes are outwardly identical, and females are not known to bite, this may serve as gender recognition behavior.) An interested female then pivots to present a rear approach, and mounting quickly follows.

Clutch size averages seven eggs, each one the size of a quarter, deposited in the sandy mound at the burrow's entrance or in another sunny location. Incubation takes one hundred days, although most eggs never hatch due to predation by foxes, raccoons, opossums, and similar varmints.

Humans have also devised ways of preying on gophers, even when they are tucked seemingly safe underground. The "pulling" of gophers is a time-honored tradition, reaching the height of its popularity during the Great Depression, when a "Hoover chicken" meant meat on the table for many southern families. A gopher puller threads a long, flexible wire into a burrow and "fishes" until the wire's terminal hook catches under an animal's lower shell. While the tortoise desperately resists— digging its forelimbs into the tunnel walls and emitting an indignant hiss—the puller undertakes a series of vigorous, protracted tugs, finally dragging his still protesting trophy to daylight. After butchering with hammer and knife, the tough and scrawny carcass is usually stewed.

Human predation, when combined with habitat fragmentation, modern intensive forestry practices, and highway mortality, has produced precipitous declines in gopher numbers—an estimated 80 percent— over the past century. This decline led the U. S. Fish and Wildlife Service in 1987 to list certain Gulf Coast populations as "threatened," each with an elaborate (and expensive) protection plan.

However it is accomplished, protection from us humans is essential for the continued survival of these trusting, mild-mannered creatures. In a very real way, their "hole" existence depends on us.

Focus your research on one of the co-inhabitants of gopher tortoise burrows—gopher frog, eastern indigo snake, or eastern diamondback rattler. How does "your" species interact with the gopher tortoise? Specifically, how do the two species share that space, and what role does each play in the longleaf pine ecosystem?

Siren intermedia—homely salamander or heartless chanteuse?
Photo by W. Mike Howell.

The Siren's Song

In nature study, it's important to notice the bizarre as well as the beautiful. And few creatures are as bizarre, at least to us, as a siren.

There's a certain romance to natural history—or, at least, there can be. And it took a truly romantic soul to stick the name "siren" on some of nature's homeliest creatures. Eel-like in shape, with slimy skin, bushy gills, tiny lidless eyes, and no hind limbs, sirens smack of hideous monsters sallying forth from the antediluvian ooze. One such species, *Siren intermedia,* inhabits quiet, shallow, often turbid waters where vegetation abounds—swamps, ditches, sloughs, ponds, and lakes. Other than an occasional much-startled fisherman, few people ever see these secretive salamanders. They spend their days buried in the mud or concealed in vegetation, emerging at night to forage on arthropods, worms, and algae. In turn, they are "taken" by aquatic predators like water moccasins.

Besides that, little is known of a siren's life history,

especially its reproductive behavior, and much guesswork is applied. Sexual maturity is apparently reached in two years, and in early spring each adult female lays a clump of about two hundred eggs in a two-inch pocket at the bottom of a pond. (Strong evidence points toward external fertilization, but no one knows for sure.) Hatchlings, half an inch long, sport prominent longitudinal stripes. Adults lose those stripes, becoming dark brown, black, dark olive, olive green, or grayish blue, often with subtle spots, attaining a final length of ten to fifteen inches.

Like other amphibians, sirens must undergo metamorphosis in order to reach adulthood, but theirs is a markedly incomplete one. In fact, two strange words describe the strange way that sirens metamorphose. Adult sirens are *neotenic* creatures, looking like permanent larvae—fully capable of reproduction but larval in form. They are also *paedomorphic,* since many of their adult characteristics are retained from larvahood—the aquatic habit, prominent gills and dorsal fin, and lack of hind legs. In fact, the only part that completely metamorphoses is their skin. But that skin becomes a major means of survival, especially in the precarious situations in which sirens too often find themselves.

Siren skin is notoriously slimy. If a pond begins to dry, already hyperactive skin glands secrete a cocoon which covers the entire body, except for the mouth. Dry, inelastic, and parchment-like in consistency, this protective sheath perfectly parallels the cocoons of African lungfishes and serves the same purpose. It retards desiccation and permits sirens to remain in usually aquatic but periodically dry environments. (With only front legs—and puny ones at that—how could they walk to a wetter spot, anyway?) By severely reducing their metabolism and using extensive fat reserves for energy, sirens can survive such entombment for at least thirty-five weeks, and probably longer.

All this, oddly enough, brings us back to the name "siren." In Greek mythology, sirens are insidious temptresses who, by their sweet singing, lure unsuspecting mariners to sure and total destruction. And it's while holed up in its temporary tomb that *Siren intermedia* is likely to burst into seductive song—well, actually, more like plaintive yelps, low whistles, subdued bleats, and assorted clicking noises. Unlike most salamanders, sirens make and respond to sounds that have communicational significance. They whimper while entombed; also, when they butt or bite each other, the displaced or injured individual swims away emitting pitiful cries, preventing further attack. Clicking sounds abound when a siren enters a strange area or leaves a burrow to gulp air at the surface. Since sirens lack vocal cords, their means of sound production remains very much a mystery, and this assortment of noises may be produced in a most unromantic way, like head-jerking associated with the clapping of horny jaw plates. The result is not exactly "The Indian Love Song" of Hollywood fame, but still an effective means of communication.

So, to a naturalist with a romantic bent, the forlorn bleats of an entombed or injured, seldom seen, ridiculously slimy, two-legged aquatic salamander become the bewitching emanations of a silken, curvaceous, mermaid-like enchantress. And why not?

On your next nature walk, look for something ugly, bizarre, or scary—an amphibian or a snake, perhaps a worm. What makes it appear so to you? Investigate the life cycle and habitat needs of this organism. How do its "ugly" features help it to survive and thrive?

_____

_____

_____

_____

_____

_____

_____

_____

_____

_____

_____

_____

_____

_____

_____

_____

_____

_____

_____

_____

_____

_____

A morel awaits its culinary fate in Jefferson County, Alabama.
Photo by W. Mike Howell.

chapter twenty-five

Lessons in Morelity

I've been chasing morels—pretty much unsuccessfully—since my undergraduate college days. They are mysterious, ephemeral, delicious, and well worth the effort.

The morel of this story is a fungus—a fungus bathed in mystery, intrigue, and even occasional death. A fungus also sautéed in olive oil, with perhaps a touch of garlic, and relished by the most discriminating epicures.

Like other fungi, these gastronomic delights grow from an unseen underground mycelium, a network of thin tubes that absorbs nutrients from the organic layer of the soil. (Connected to tree roots, the mycelium helps to "feed" those trees, the two completely unrelated creatures forming an intimate and vital association.) Such a network remains unnoticed until its fruiting bodies pop up through the soil surface to spread their spores, the fissured, puckered, and honey-combed caps appearing as golden brown brains-on-sticks. Deer, squirrels, and other four-legged gourmands gorge on the succulent caps, which concentrate thirty to one hundred times the basic minerals normally found in everyday foliage. (A traditional morel value?)

Thirty species of morels (*Morchella*) occur worldwide, with several in the southeastern United States. Here they occupy a variety of habitats, from deciduous woods to apple orchards and suburban lawns; according to folklore, they appear in the spring "when oak leaves are the size of mouse ears." Their sporadic and unpredictable appearance adds much lure to such lore. One ultra-serious mycophile takes a ridge-and-valley approach, swearing that the best ones reside halfway up the northeast face of a ridge running due southeast-northwest; always start at the bottom and work your way up. (Taking the morel high ground?) But what if you stumble on just one lonely "bloom"? To pick (or not to pick) becomes a true morel dilemma.

Morel indifference is rare. Instead, an absolute mania surrounds the 'shrooms. Seasoned hunters hide their "honey spot," even from close friends and family; some bear that secret to the grave. (It's always best to learn from such experienced folks. As with all mushrooms, utmost care must be taken in correctly identifying your dinner because imposters and look-a-likes abound, and any a-morel acts can prove deadly.) Rollo Leach, a life-long New England mycophagist/poet, offered this insightful verse:

He who secret reveals

And nothing conceals

Somehow never tells

Where grow morels.

Occasionally, morelists enjoy banner years with bumper crops. When the soil, rain, and temperature meet in precise, cosmic perfection, fungi "flush" in incredible numbers, creating a "carpet" of pot-of-gold proportions. This phenomenon is especially pronounced in the Pacific Northwest and Alaska, where the morel majority burst forth in recently burned areas, especially during the first warm wet spring after a fire. Folks quit their day-jobs to wander those charbroiled forests, fetching fifteen to twenty dollars per pound in quick, no-questions-asked cash. (Of course, as with any commodity, the price quickly plummets when supply outpaces demand.) And as with any gold rush, unsavory types also seek the savory bounty, resulting in turf battles, claim jumping, fist fights, and occasional murders. Old-timers treat newcomers with great suspicion and open contempt, accusing greenhorns of destructive, non-renewable, unsustainable collecting practices and other im-morel acts.

But let's put all that aside for now. However you obtained your exquisite, esculent morsels, wash them well and pat them dry. First a dip in an egg wash, then a gentle roll in personally seasoned bread crumbs. Next sauté in olive oil, drain, and enjoy. (The unique taste is variously described as meaty, musty, crab-like, clam-like, and "like spring in your mouth.") After all, it's your morel obligation.

Choose another epicurean fungus, like the chanterelle, and investigate its life history. Which host plants does it demand, and which habitat does it prefer? What about climate conditions for its "flushing"? Where and when might you best find them in your region?

_____

_____

_____

_____

_____

_____

_____

_____

_____

_____

_____

_____

_____

_____

_____

_____

_____

_____

_____

_____

_____

_____

_____

_____

Further Reading

INTRODUCTION

The following works give you the "how to" of observing
nature, keeping a nature journal, and creating your own essays.
Appropriately enough, the first one comes from the Golden
Age of Nature Study (as I have dubbed the years 1880 to 1920).
Keeping a Nature Journal teaches you how to literally draw
from nature, something at which I'm not skilled. E. O. Wilson's
book is highly recommended—the autobiography of the world's
premier naturalist, born and raised in Alabama.

Comstock, Anna Botsford. *Handbook of Nature-Study*. Ithaca,
NY: Cornell University Press, 1911.

Hancock, Elise. *Ideas into Words: Mastering the Craft of Science
Writing*. Baltimore, MD: Johns Hopkins University Press,
2003.

Leslie, Clare Walker, and Charles E. Roth. *Keeping a Nature
Journal: Discover a Whole New Way of Seeing the World
Around You*. 2nd ed. North Adams, MA: Storey Publishing,
2003.

Murray, John A. *Writing About Nature: A Creative Guide*. Rev.
ed. Albuquerque: University of New Mexico Press, 2003.

Wilson, Edward O. *Naturalist*. Washington: Island Press, 1994.

CHAPTER 1

Full participation pieces, such as this one, demand little advance preparation—just a good attitude. For three days I was in the presence of experts, who carried most of their knowledge of the plants and animals of the Walls of Jericho in their heads. I just took notes on what occurred and what was found.

This is where a field notebook is essential. It's always best to capture the sights, sounds, and interactions of such an adventure as they are experienced. Then it's just a matter of fusing your musings into a coherent whole. My notebook from that three-day weekend contains all of the essential lines, stories, and themes of the final piece.

To shore up the Davy Crockett theme—and to make sure I was being true to his spirit—I read his autobiography. And I made sure to include some of his quaint expressions. It felt good to walk in his footsteps (and talk his talk)!

Crockett, David. *A Narrative of the Life of David Crockett, of the State of Tennessee, Written by Himself.* Philadelphia, PA: E. L. Carey and A. Hart; Baltimore, MD: Carey, Hart, and Co., 1834.

Chapter 2

I always enjoy being in the field with experts, people who have devoted their lives to nature study and who are intent on passing that knowledge on to others. They have a passion that just isn't equaled by ordinary folks. I take any chance I can to walk in their woods and seine in their streams.

It took some serious library work, but I traced W. Mike Howell's story on the naming of this fish to the classic work by Jordan and Evermann listed below; another portion was quoted in the essay itself. For the details of the fish's distribution, life cycle, and life style, I turned to several recent books on darters and the fishes of the United States.

Boshung, Herbert T., Jr., and Richard L. Mayden. *Fishes of Alabama*. Washington, D.C.: Smithsonian Books, 2004.

Etnier, David A., and Wayne C. Starnes. *The Fishes of Tennessee*. Knoxville: University of Tennessee Press, 1993.

Jenkins, Robert E., and Noel M. Burkhead. *Freshwater Fishes of Virginia*. Bethesda, MD: American Fisheries Society, 1994.

Jordan, David Starr, and Barton W. Evermann. *The Fishes of North and Middle America*. 1923. Reprint, Jersey City, NJ: TFH Publications, 1963.

Kuehne, Robert A., and Roger W. Barbour. *The American Darters*. Lexington: University Press of Kentucky, 1983.

Mettee, Maurice F., Patrick E. O'Neil, and J. Malcolm Pierson. *Fishes of Alabama and the Mobile Basin*. Birmingham, AL: Oxmoor House, 1996.

Page, Lawrence M. *Handbook of Darters*. Neptune City, NY: TFH Publications, 1983.

Robison, Henry W., and Thomas M. Buchanan. *Fishes of Arkansas*. Fayetteville: University of Arkansas Press, 1988.

CHAPTER 3

All writers have stories that tug at them, that they just have to tell. And this is one of mine. It was also a difficult piece to write because, in broadcasting that childhood memory, I revealed a lot about myself—which is not something that a scientist is wont to do!

Thanks to John James Audubon, Americans have long been fascinated by birds—their beauty, behavior, and biogeography. I turned to national, regional, and state field guides for general information, then consulted the more specific volumes listed below. (Arthur Bent's life history series is a particularly valuable source of information on birds and their behaviors.) Because cranes and herons are so large and vulnerable, numerous popular articles *(Smithsonian)* bring their stories to the American public.

Bent, Arthur C. *Life Histories of North American Marsh Birds.* 1926. Reprint, New York: Dover Publications, 1963.

Burton, Robert. *Bird Behavior.* New York: Knopf/Random House, 1985.

Butler, Robert W. *The Great Blue Heron: A Natural History and Ecology of a Seashore Sentinel.* Vancouver: University of British Columbia Press, 1997.

Forbush, Edward H., and John B. May. *A Natural History of American Birds of Eastern and Central North America.* New York: Bramhall House, 1955.

Hancock, James, and James Kushlan. *The Herons Handbook.* New York: Harper & Row, 1984.

Horton, Tom. "Great Blues Are Going Great Guns." *Smithsonian* 29 (April 1999): 130–140.

Imhof, Thomas A. *Alabama Birds.* 2nd ed. Tuscaloosa: University of Alabama Press, 1976.

Chapter 4

I used Internet sources almost exclusively in researching this piece. Why? Because doodlebugs and ant lions make a global topic, and I was most interested in the personal side—the stories that people feel compelled to share with each other. The site http://www.antlionpit.com was particularly helpful, supplying everything I needed—anecdotes, natural history, and even literary references. Using Internet sites is risky, however, due to the large amount of contradictory and downright wrong information that is freely shared. I encourage you to follow my lead: Check any Internet "facts" against a published scientific standard before you accept them to be true.

CHAPTER 5

A topic like this one—an in-depth look at the life history of a single species—requires a healthy amount of basic library research. Field guides and natural history volumes provide the basic facts, but the most specific and up-to-date details are found only in scholarly journals in that particular field. For American reptiles, amphibians, and fishes, I always consult *Copeia* and *Ecology*.

Barbour, Roger W. *Amphibians and Reptiles of Kentucky.* Lexington: University Press of Kentucky, 1971.

Conant, Roger. *A Field Guide to Reptiles and Amphibians of the United States and Canada East of the 100th Meridian.* Peterson Field Guide Series, 12. Boston: Houghton Mifflin, 1958.

Cooper, William E., Jr., and Nina Burns. "Social Significance of Ventrolateral Coloration in the Fence Lizard *(Seeloporus undulatus)*." *Animal Behavior* 35 (1987): 526–532.

Crenshaw, John W., Jr. "The Life History of the Southern Spiny Lizard, *Sceloporus undulatus undulatus* Latreille." *American Midland Naturalist* 54 (1955): 257–298.

Ditmars, Raymond L. *The Reptile Book: A Comprehensive, Popularised Work on the Structure and Habits of the Turtles, Tortoises, Crocodilians, Lizards and Snakes Which Inhabit the United States and Northern Mexico.* New York: Double, Page & Company, 1908.

Dundee, Harold A., and Douglas A. Rossman. *The Amphibians and Reptiles of Louisiana.* Baton Rouge: Louisiana State University Press, 1989.

Martof, Bernard S. *Amphibians and Reptiles of Georgia: A Guide.* Athens: University of Georgia Press, 1956.

Mount, Robert H. *The Reptiles and Amphibians of Alabama.* 1975. Reprint, Tuscaloosa: University of Alabama Press, 1996.

Parker, William S. "Demography of the Fence Lizard, *Sceloporus undulatus,* in Northern Mississippi." *Copeia* 1994: 136–152.

Smith, Hobart M. *Handbook of Lizards: Lizards of the United States and of Canada.* Ithaca, NY: Comstock Publishing, 1946.

Tinkle, Donald W., and Royce E. Ballinger. "*Sceloporus undulatus:* A Study of the Intraspecific Comparative Demography of a Lizard." *Ecology* 53 (1972): 570–584.

Tinkle, Donald W., and Arthur E. Dunham. "Comparative Life Histories of Two Syntopic Sceloporine Lizards." *Copeia* 1986: 1–18.

CHAPTER 6

Few insects have as much folklore surrounding them as dragonflies and damselflies. Plus, their life cycles are truly fantastic. Information on insect life cycles is best found in field guides, entomology textbooks, and natural history tomes. The oldest such tomes are often the best sources because they are filled with unabashed "golly gee" observations and analogies.

Bick, J. C., and D. Sulzbach. "Reproductive Behaviour of the Damselfly, *Heta-erina americana* (Fabricius) (Odonata: Calopterygidae)." *Animal Behaviour* 14 (1966): 156–158.

Clausen, Lucy W. *Insect Fact and Folklore.* New York: Macmillan, 1954.

Comstock, Anna Botsford. *Handbook of Nature-Study.* Ithaca, NY: Cornell University Press, 1911.

Comstock, John H. *An Introduction to Entomology.* 9th rev. ed. Ithaca, NY: Comstock Publishing, 1940.

Comstock, John H., and Anna Botsford Comstock. *A Manual of the Study of Insects.* 15th ed. Ithaca, NY: Comstock Publishing, 1917.

Corbet, P. S. "Biology of Odonata." *Annual Review of Entomology* 25 (1980): 189–217.

Dunkle, S. W. *Damselflies of Florida, Bermuda, and the Bahamas.* Enfield, NH: Scientific Publishers, 1990.

Gillott, Cedric. *Entomology.* New York: Plenum Press, 1980.

Imms, A. D. *A General Textbook of Entomology, Including the Anatomy, Physiology, Development, and Classification of Insects.* 7th ed. New York: E. P. Dutton, 1948.

Matthews, Robert W., and Janice R. Matthews. *Insect Behavior.* New York: Wiley, 1978.

Waldbauer, Gilbert. *Insects Through the Seasons.* Cambridge, MA: Harvard University Press, 1996.

Chapter 7

Field guides are the best sources of information for a topic like this one, pulling together all of the pertinent information on a butterfly's size, color forms, distribution, and host plants. Fortunately, guides are available for every region of the United States and most countries.

Due to their beauty, grace, and intricate life cycles, butterflies are prime topics for natural history books as well. And because of their declining numbers—related to the loss of host plants and their habitats—many popular articles artistically describe their precarious situation.

Ehrlich, Paul R., and Anne H. Ehrlich. *How to Know the Butterflies.* Dubuque, IA: W. C. Brown, 1961.

Holland, W. J. *The Butterfly Book: A Popular Guide to a Knowledge of the Butterflies of North America.* Garden City, NY: Doubleday, Page & Co., 1914.

Howe, William H. *The Butterflies of North America*. Garden City, NY: Doubleday, 1975.

Klots, Alexander B. *A Field Guide to the Butterflies of North America, East of
the Great Plains*. Peterson Field Guide Series, 4. Boston: Houghton Mifflin,
1951.

Scott, James A. *The Butterflies of North America: A Natural History and Field
Guide*. Stanford, CA: Stanford University Press, 1986.

CHAPTER 8

Being nocturnal, moths are not noticed (or worried about) as much as butterflies. But they show equivalent levels of variation and dependence on specific
host plants. So again, field guides and natural history volumes provide the basic
information on their life histories and variety.

Much has been written about humans as truly biological creatures and our
inability to "escape" that biology. *The Naked Ape,* listed below, is rather dated
but a classic in this field.

Brown, Richard E., and David W. Macdonald, eds. *Social Odours in Mammals*.
London: Oxford University Press, 1985.

Forsyth, Adrian. *A Natural History of Sex: The Ecology and Evolution of Mating
Behavior*. New York: Scribner's, 1986.

Holland, W. J. *The Moth Book: A Popular Guide to a Knowledge of the Moths of
North America*. New York: Doubleday, Page & Co., 1905.

McClintock, Martha K. "Menstrual Synchrony and Suppression." *Nature* 229
(1971): 244–245.

Morris, Desmond. *The Naked Ape*. New York, McGraw-Hill, 1967.

Tuskes, Paul M., James P. Tuttle, and Michael M. Collins. *The Wild Silk Moths of North America: A Natural History of the Saturniidae of the United States and Canada.* Ithaca, NY: Cornell University Press, 1996.

Wilson, Edward O. *On Human Nature.* Cambridge, MA: Harvard University Press, 1978.

CHAPTER 9

I was lucky to be around while W. Mike Howell and Ronald Jenkins were working on their spider book (see below). It became a communal effort, as their students and colleagues shared in bringing in specimens and stretching the distribution records. One colleague, Carl Ponder, brought in a bolas spider from Cullman County, Alabama, and I knew I had to write about it.

Although this piece is obviously based on the Hutchinson article, I also consulted recent works in systematic and chemical ecology. References to major works of the Golden Age of Nature Study can be found in the endnotes of preceding and succeeding chapters.

Eberhard, William G. "Aggressive Chemical Mimicry by a Bolas Spider." *Science* 198 (1977): 1173–1175.

———. "The Natural History and Behavior of the Bolas Spider *Mastophora dizzydeani* sp. n. (Araneidae)." *Psyche* 87 (1980): 143–169.

Gertsch, Willis J. "The North American Bolas Spiders of the Genera *Mastophora* and *Agatostichus*." *Bulletin of the American Museum of Natural History* 106 (1955): 221–254.

Haynes, K. F., C. Gemeno, K. V. Yeargan, J. G. Millar, and K. M. Johnson. "Aggressive Chemical Mimicry of Moth Pheromones by a Bolas Spider: How Does This Specialist Predator Attract More Than One Species of Prey?" *Chemoecology* 12 (2002): 99–105.

Howell, W. Mike, and Ronald L. Jenkins. *Spiders of the Eastern United States: A Photographic Guide.* Boston: Pearson Education, 2004.

Hutchinson, Charles E. "A Bolas-Throwing Spider." *Scientific American* 89 (1903): 172.

Stowe, Mark K., James H. Tumlinson, and Robert R. Heath. "Chemical Mimicry: Bolas Spiders Emit Components of Moth Prey Species Sex Pheromones." *Science* 236 (1987): 964–967.

Yeargan, Kenneth V. "Biology of Bolas Spiders." *Annual Review of Entomology* 39 (1994): 81–99.

———. "Ecology of a Bolas Spider, *Mastophora hutchinsoni:* Phenology, Hunting Tactics, and Evidence for Aggressive Chemical Mimicry." *Oecologia* 74 (1988): 524–530.

CHAPTER 10

Animal pollination involves the intricate and intimate relationship between plants and their pollinators. Biologists speak of the coevolution of the two groups, with modifications of one being met and matched by the other. Over evolutionary time, strange and wonderful coadaptations have been made—in some cases, a plant species is pollinated only by a single animal species, with the two completely dependent on each other.

Orchid pollination is particularly bizarre, and biologists like Charles Darwin have devoted years (and entire books) to its study. (It's interesting to note, and probably not an accident, that Darwin chose this topic for his first book after *The Origin of Species.*) I include those references in the list below, as well as several natural history books and field guides.

Reference was made in this piece to *Bumblebee Economics,* by the New England naturalist Bernd Heinrich. I encourage you to examine his excellent, thoughtful works.

Bentley, Stanley L. *Native Orchids of the Southern Appalachian Mountains.* Chapel Hill: University of North Carolina Press, 2000.

Correll, Donovan S. *Native Orchids of North America North of Mexico.* Waltham, MA: Chronica Botanica, 1950.

Darwin, Charles. *On the Various Contrivances by which British and Foreign Orchids Are Fertilised by Insects, and on the Good Effects of Intercrossing.* London: John Murray, 1862.

Davis, Richard W. "The Pollination Biology of *Cypripedium acaule* (Orchidaceae)." *Rhodora* 88 (1986): 445–450.

Dodson, Calaway H. "Studies in Orchid Pollination: *Cypripedium, Phragmopedium* and Allied Genera." *Bulletin of the American Orchid Society* 35 (1966): 125–128.

Dressler, Robert L. *The Orchids: Natural History and Classification.* Cambridge, MA: Harvard University Press, 1981.

Heinrich, Bernd. *Bumblebee Economics.* Cambridge, MA: Harvard University Press, 1979.

Luer, Carl A. *The Native Orchids of the United States and Canada, Excluding Florida.* Bronx, NY: New York Botanical Garden, 1975.

Stoutamire, Warren P. "Flower Biology of the Lady's Slippers (Orchidaceae: *Cypripedium*)." *Michigan Botanist* 6 (1967): 159–175.

CHAPTER 11

Properly calculating reproductive costs—whether instinctively or actually—is basic to a species' survival. The jack-in-the-pulpit sex change strategy allows just that, even though an individual might "suffer" for a season or two.

The basic references for this piece were gathered by my student, Carolee Franklin, for a field study that she conducted near the Samford University campus. I also consulted various technical papers on sex choice and change.

Bierzychudek, Pauline. "Assessing 'Optimal' Life Histories in a Fluctuating Environment: The Evolution of Sex-Changing by Jack-in-the-Pulpit." *American Naturalist* 123 (1984): 829–840.

———. "The Demography of Jack-in-the-Pulpit, a Forest Perennial that Changes Sex." *Ecological Monographs* 52 (1982): 335–351.

———. "Determinants of Gender in Jack-in-the-Pulpit: The Influence of Plant Size and Reproductive History." *Oecologia* 65 (1984): 14–18.

Clay, Keith. "Size-Dependent Gender Change in Green Dragon (*Arisaema dracontium;* Araceae)." *American Journal of Botany* 80 (1993): 769–777.

Freeman, D. C., K. T. Harper, and E. L. Charnov. "Sex Change in Plants: Old and New Observations and New Hypotheses." *Oecologia* 47 (1980): 222–232.

Leigh, Egbert G., Jr., Eric L. Charnov, and Robert R. Warner. "Sex Ratio, Sex Change, and Natural Selection." *Proceedings of the National Academy of Sciences* 73 (1976): 3656–3660.

Lovett Doust, Jon, and Paul B. Cavers. "Sex and Gender Dynamics in Jack-in-the-Pulpit, *Arisaema triphyllum* (Araceae)." *Ecology* 63 (1982): 797–808.

Lovett Doust, Lesley, Jon Lovett Doust, and Karen Turi. "Fecundity and Size Relationships in Jack-in-the-Pulpit, *Arisaema triphyllum* (Araceae)." *American Journal of Botany* 73 (1986): 489–494.

Policansky, David. "Sex Choice and Reproductive Costs in Jack-in-the-Pulpit." *BioScience* 37 (1987): 476–481.

———. "Sex Choice and the Size Advantage Model in Jack-in-the-Pulpit (*Arisaema triphyllum*)." *Proceedings of the National Academy of Sciences* 78 (1981): 1306–1308.

Rust, Richard W. "Pollen Movement and Reproduction in *Arisaema triphyllum.*" *Bulletin of the Torrey Botanical Club* 107 (1980): 539–542.

Sakamoto, Sadao. "*Arisaema triphyllum,* Jack-in-the-Pulpit, in Minnesota,

Especially at the Cedar Creek Natural History Area." *Proceedings of the Minnesota Academy of Science* 29 (1961): 153–168.

Chapter 12

Field guides make a good starting point for a piece like this. And because people are so fascinated by these timeless migration events, lots of good popular articles are available.

I also encourage you to join Gulf Coast birding events, held each fall and spring, or make your own event by following the Alabama Coastal Birding Trail.

Porter, John, Jr. *A Birder's Guide to Alabama.* Tuscaloosa: University of Alabama Press, 2001.

Porter, John, Jr., and Jackie Porter, eds. *The Alabama Coastal Birding Trail.* Daphne, AL: U.S. Fish & Wildlife Service, 2001.

Weidensaul, Scott. "Across the Gulf on a Wing and a Prayer." *Nature Conservancy* 54 (2004): 34–41.

Chapter 13

Few field guides are written for the fossils. Instead, such information must be gleaned from technical manuals and geology treatments. The ones that I used for this chapter are listed here.

Adams, George I., Charles Butts, L. W. Stephenson, and Wythe Cooke. *Special Report 14: Geology of Alabama.* Tuscaloosa: Geological Survey of Alabama, 1926.

Lacefield, Jim. *Lost Worlds in Alabama Rocks.* Tuscaloosa: Alabama Geological Society, 2000.

Thurmond, John T., and Douglas E. Jones. *Fossil Vertebrates of Alabama.* Tuscaloosa: University of Alabama Press, 1981.

CHAPTER 14

The southeastern states, and especially Alabama, have long been known for their diversity of aquatic organisms, including freshwater mussels. Unfortunately, much of that diversity must now be described in the past tense, because the building of dams (for hydroelectricity and navigation) condemned many species to extinction. We have good records, though, in the technical papers of malacology. The two recent treatments for Tennessee and Alabama, listed below, are truly beautiful works that pull together all that is known about the mussels of these two states.

Haag, Wendell R., Robert S. Butler, and Paul D. Hartfield. "An Extraordinary Reproductive Strategy in Freshwater Bivalves: Prey Mimicry to Facilitate Larval Dispersal." *Freshwater Biology* 34 (1995): 471–476.

Haag, Wendell R., and Melvin L. Warren, Jr. "Host Fishes and Reproductive Biology of Six Freshwater Mussel Species from the Mobile Basin, USA." *Journal of the North American Benthological Society* 16 (1997): 576–585.

Haag, Wendell R., Melvin L. Warren Jr., and Mahala Shillingsford. "Host Fishes and Host-Attracting Behavior of *Lampsilis altilis* and *Villosa vibex* (Bivalvia: Unionidae)." *American Midland Naturalist* 141 (1999): 149–157.

Jones, Robert L., William T. Slack, and Paul D. Hartfield. "The Freshwater Mussels (Mollusca: Bivalvia: Unionidae) of Mississippi." *Southeastern Naturalist* 4 (2005): 77–92.

McGregor, Stuart W., Patrick E. O'Neil, and J. Malcolm Pierson. "Status of the Freshwater Mussel (Bivalvia: Unionidae) Fauna in the Cahaba River System, Alabama." *Walkerana* 11 (2000): 215–237.

Parmalee, Paul W., and Arthur E. Bogan. *The Freshwater Mussels of Tennessee.* Knoxville: University of Tennessee Press, 1998.

Pilarczyk, Megan M., Paul M. Stewart, Douglas N. Shelton, Holly N. Blalock-Herod, and James D. Williams. "Current and Recent Historical Freshwater Mussel Assemblages in the Gulf Coastal Plains." *Southeastern Naturalist* 5 (2006): 205–226.

van der Shalie, Henry. "Mollusks in the Alabama River Drainage: Past and Present." *Sterkiana* 71 (1981): 24–40.

———. *Occasional Papers of the Museum of Zoology 392: The Naiades (Fresh-Water Mussels) of the Cahaba River in Northern Alabama.* Ann Arbor: University of Michigan, 1938.

Williams, James D., Arthur E. Bogan, and Jeffrey T. Garner. *Freshwater Mussels of Alabama and the Mobile Basin in Georgia, Mississippi, and Tennessee.* Tuscaloosa: University of Alabama Press, 2008.

CHAPTER 15

Every thirteen years—*precisely* every thirteen years—people in the American South get excited about periodical cicadas. When it happens in your area, there will be lots of information on the Internet about cicada life cycles and sightings. More technical information can be gleaned from entomological and ecological journals, including those listed below.

May, Robert M. "Periodical Cicadas." *Nature* 277 (1979): 347–349.

Williams, Kathy S., and Chris Simon. "The Ecology, Behavior, and Evolution of Periodical Cicadas." *Annual Review of Entomology* 40 (1995): 269–295.

Williams, Kathy S., Kimberly G. Smith, and Frederick M. Stephen. "Emergence of 13-Yr Periodical Cicadas (Cicadidae: *Magicicada*): Phenology, Mortality, and Predator Satiation." *Ecology* 74 (1993): 1143–1152.

Yoshimura, Jin. "The Evolutionary Origins of Periodical Cicadas during Ice Ages." *American Naturalist* 149 (1997): 112–124.

Chapter 16

We humans, being mammals, have a special affinity for our brother and sister species. So there are plenty of good mammal books available, with several listed below. The Gosse book has an interesting description of an Alabama possum hunt during the 1830s; the Audubon and Bachman tome is worth perusing for its beautiful artwork alone. I've also included a book on tropical mammals for those who want to venture into those regions.

Audubon, John James, and John Bachman. *The Viviparous Quadrupeds of North America.* New York: V. G. Audubon, 1854. (Several reprints are available.)

Emmons, Louise H. *Neotropical Rainforest Mammals: A Field Guide.* 2nd ed. Chicago: University of Chicago Press, 1997.

Gosse, Philip Henry. *Letters from Alabama, (U.S.) Chiefly Relating to Natural History.* London: Morgan and Chase, 1859. Reprint, Tuscaloosa: University of Alabama Press, 1993.

Hall, E. Raymond. *The Mammals of North America.* New York: Wiley, 1981.

Whitaker, John O., Jr., and William J. Hamilton Jr. *Mammals of the Eastern United States.* Ithaca, NY: Comstock Publishing, 1998.

Chapter 17

There are lots of good sources of information on endangered bird species—the Internet, field guides, ornithology journals, and natural history texts. I also consulted the *Federal Register* for rulings on the protection of brown pelicans.

Bailey, Alfred M. "The Brown Pelican—a Good Citizen." *Wilson Bulletin* 30 (1918): 65–68.

———. "The Brown Pelicans: A Series of Previously Unpublished Photographs of the Breeding Birds on the Louisiana Gulf Coast, with Notes on Their Haunts and Habits." *Natural History* 20 (1920): 197–201.

Bent, Arthur C. *Life Histories of North American Petrels and Pelicans and Their Allies.* 1922. Reprint, New York: Dover Publications, 1964.

Forbush, Edward H., and John B. May. *A Natural History of American Birds of Eastern and Central North America.* New York, Bramhall House, 1955.

Imhof, Thomas A. *Alabama Birds.* 2nd ed. Tuscaloosa: University of Alabama Press, 1976.

Lowery, George H. *Louisiana Birds.* Rev. 2nd ed. Baton Rouge: Louisiana State University Press, 1960.

McNease, Larry, Ted Joanen, David Richard, Joseph Shepard, and Stephen A. Nesbitt. "The Brown Pelican Restocking Program in Louisiana." *Proceedings of the Annual Conference, Southeastern Association of Fish and Wildlife Agencies* 38 (1984): 165–173.

U.S. Fish and Wildlife Service. "Endangered and Threatened Wildlife and Plants: Removal of the Brown Pelican in the Southeastern United States from the List of Endangered and Threatened Wildlife." *Federal Register* 50 (1985): 4938–4945.

Wilkinson, Philip M., Stephen A. Nesbitt, and James F. Parnell. "Recent History and Status of the Eastern Brown Pelican." *Wildlife Society Bulletin* 22 (1994): 420–430.

CHAPTER 18

Because of its widespread occurrence and odd appearance, Spanish moss is described, depicted, and discussed a great deal in both scientific literature and the popular press. The Internet is also full of information and anecdotes. To catch a fuller flavor, I consulted several classic works on Southeastern natural history, including Bartram, Gosse, Gray, and Rafinesque. I also looked at recent papers on natural products extracted from Spanish moss.

Bartram, William. *Travels Through North & South Carolina, Georgia, East & West Florida, the Cherokee Country, the Extensive Territories of the Muscogulges, or Creek Confederacy, and the Country of the Chactaws; Containing an Account of the Soil and Natural Productions of those Regions, Together with Observations on the Manners of the Indians.* Philadelphia: James & Johnson, 1791. (Many reprints are available; Francis Harper's 1958 Naturalist's Edition is highly recommended.)

Benzing, David H., ed. *Bromeliaceae: Profile of an Adaptive Radiation.* London: Cambridge University Press, 2000.

Duncan, Wilbur H., and Leonard E. Foote. *Wildflowers of the Southeastern United States.* Athens: University of Georgia Press, 1975.

Gosse, Philip Henry. *Letters from Alabama, (U.S.) Chiefly Relating to Natural History.* London: Morgan and Chase, 1859. Reprint, Tuscaloosa: University of Alabama Press, 1993.

Gray, Asa, and L. H. Bailey. *Field, Forest and Garden Botany: A Simple Introduction to the Common Plants of the United States East of the 100th Meridian, both Wild and Cultivated.* New York: American Book Company, 1895.

Greene, Wilhelmina F., and Hugo L. Blomquist. *Flowers of the South, Native and Exotic.* Chapel Hill: University of North Carolina Press, 1953.

Rafinesque, C. S. *Florula Ludoviciana; or, A Flora of the State of Louisiana.* 1817. Reprint, New York: Hafner, 1967.

Witherup, K. M., J. L. McLaughlin, R. L. Judd, M. H. Ziegler, P. J. Medon, and

W. J. Keller. "Identification of 3-Hydroxy-3-Methylglutaric Acid (HMG) as a Hypoglycemic Principle of Spanish Moss *(Tillandsia usneoides)." Journal of Natural Products* 58 (1995): 1285–1290.

CHAPTER 19

This chapter comes directly from a visit to the Estuarium on Dauphin Island, where I first came face to distorted face with a living flounder—which changed, forever, the way I view seafood.

Information on flatfishes is found in many sources, from natural history books to physiology texts to cookbooks. I found Bob Shipp's guide to be most helpful and co-opted some of his clever phrasings.

Boschung, Herbert T. *Bulletin 14: Catalog of Freshwater and Marine Fishes of Alabama.* Tuscaloosa: Alabama Museum of Natural History, 1992.

Boschung, Herbert T., and Richard L. Mayden. *Fishes of Alabama.* Washington, DC: Smithsonian Institution, 2004.

Evans, David H., ed. *The Physiology of Fishes.* Boca Raton, FL: CRC Press, 1993.

LaGorce, John Oliver, ed. *The Book of Fishes; Revised and Enlarged Edition, Presenting the Better Known Species of Food and Game Fishes of the Coastal and Inland Waters of the United States.* Washington, DC: National Geographic Society, 1939.

Mettee, Maurice F., Patrick E. O'Neil, and J. Malcolm Pierson. *Fishes of Alabama and the Mobile Basin.* Birmingham, AL: Oxmoor House, 1996.

Shipp, Robert L. *Dr. Bob Shipp's Guide to Fishes of the Gulf of Mexico.* Mobile, AL: KME Seabooks, 1986.

CHAPTER 20

This topic is a particularly demanding one because it requires in-depth sleuthing in several areas of study, including basic botany, herbal treatments, philosophy, and the history of medicine. Because the Doctrine of Signatures is so universally embraced, there's no shortage of information, misinformation, and viewpoints to sift through.

Culpeper's book is readily available, since it has recently been reprinted by The University of Alabama Press. The history of herbals is well covered by both Anderson and Arber. A number of nineteenth-century treatments of medicinal plants provide the flavor for their uses, while the listed histories of medicine place all of this in proper perspective.

Anderson, Frank J. *An Illustrated History of the Herbals.* New York: Columbia University Press, 1977.

Arber, Agnes. *Herbals, Their Origin and Evolution; a Chapter in the History of Botany, 1470–1670.* 2nd ed. London: Cambridge University Press, 1953.

Bennett, Bradley C. "Doctrine of Signatures: An Explanation of Medicinal Plant Discovery or Dissemination of Knowledge?" *Economic Botany* 61 (2007): 246–255.

Berman, Alex, and Michael A. Flannery. *America's Botanico-Medical Movements: Vox Populi.* Binghamton, NY: Pharmaceutical Products Press, 2001.

Blackwell, Will H. *Poisonous and Medicinal Plants.* Englewood Cliffs, NJ: Prentice Hall, 1990.

Clapp, A. "A Synopsis; or, Systematic Catalogue of the Indigenous and Naturalized, Flowering and Filicoid (Exogens, Endogens, and Acrogens), Medicinal Plants of the United States, with their Localities, Botanical and Medicinal References, and a Short Account of their Medicinal Properties; Being a Report of the Committee on Indigenous Medical Botany and Material Medica for 1850–51." *Transactions of the American Medical Association* 5 (1852): 689–906.

Culpeper, Nicholas. *The English Physitian: Or an Astrologo-Physical Discourse of the Vulgar Herbs of This Nation.* London: Peter Cole, 1652. Reprint of edited and annotated 1708 American version, with introduction and footnotes by Michael A. Flannery. Tuscaloosa: University of Alabama Press, 2007.

Duke, James A. *CRC Handbook of Medicinal Herbs.* Boca Raton, FL: CRC Press, 1985.

Foster, Steven, and James A. Duke. *A Field Guide to Medicinal Plants, Eastern and Central North America.* Peterson Field Guide Series, 40. Boston: Houghton Mifflin, 1990.

Hale, Josiah. "Report on the Medical Botany of the State of Louisiana." *New Orleans Medical and Surgical Journal* 9 (1852): 152–173, 287–313.

Harris, Ben Charles. *The Compleat Herbal: Being a Description of the Origins, the Lore, the Characteristics, the Types, and the Prescribed Uses of Medicinal Herbs, Including an Alphabetical Guide to All Common Medicinal Plants.* Barre, MA: Barre Publishers, 1972.

Holley, Howard L. *The History of Medicine in Alabama.* Birmingham: University of Alabama School of Medicine, 1982.

Lewis, Walter H., and Memory P. F. Elvin-Lewis. *Medical Botany: Plants Affecting Human Health.* 2nd ed. Hoboken, NJ: John Wiley & Sons, 2003.

Lovejoy, Arthur O. *The Great Chain of Being: A Study of the History of an Idea.* Cambridge, MA: Harvard University Press, 1936.

Martin, Laura C. *Wildflower Folklore.* Chester, CT: Globe Pequot Press, 1984.

Mohr, Charles T. "The Medicinal Plants of Alabama; Systematic List of the Medicinal Plants Occurring within the Limits of the State, with Notes on Their Distribution and Proper Time of Collecting the Parts Used." *Proceedings of the Annual Meeting of the Alabama Pharmaceutical Association* 9: (1890): 45–61.

Porcher, Francis Peyre. *Resources of the Southern Fields and Forests, Medical,*

Economical, and Agricultural; Being also a Medical Botany of the Confederate States, with Practical Information on the Useful Properties of the Trees, Plants, and Shrubs. Charleston, SC: Evans & Cogswell, 1863.

CHAPTER 21

I consulted mainly field guides and the ecological literature in researching this piece. But with frogs (as with birds), there's an additional source—published tape recordings of their calls. By studying and memorizing those calls, you'll be much more "in tune" with nature.

Barbour, Roger W. *Amphibians and Reptiles of Kentucky.* Lexington: University Press of Kentucky, 1971.

Behler, John L., and F. Wayne King. *National Audubon Society Field Guide to North American Reptiles and Amphibians.* New York: Alfred A. Knopf, 1979.

Cochran, Doris M., and Coleman J. Goin. *The New Field Book of Reptiles and Amphibians.* New York: Putnam, 1970.

Conant, Roger. *A Field Guide to Reptiles and Amphibians of the United States and Canada East of the 100th Meridian.* Peterson Field Guide Series, 12. Boston: Houghton Mifflin, 1958.

Duellman, William E., and Linda Trueb. *Biology of Amphibians.* New York: McGraw Hill, 1986.

Dundee, Harold A., and Douglas A. Rossman. *The Amphibians and Reptiles of Louisiana.* Baton Rouge: Louisiana State University Press, 1989.

Elliott, Lang. *The Calls of Frogs and Toads.* Mechanicsburg, PA: Stackpole Books, 2004.

Freed, Arthur N. "Prey Selection and Feeding Behavior of the Green Treefrog (*Hyla cinerea*)." *Ecology* 61 (1980): 461–465.

Gerhardt, H. Carl, Sheldon I. Guttman, and Alvan A. Karlin. "Natural Hybrids between *Hyla cinerea* and *Hyla gratiosa:* Morphology, Vocalization and Electrophoretic Analysis." *Copeia* 1980: 577–584.

Martof, Bernard S. *Amphibians and Reptiles of Georgia: A Guide.* Athens: University of Georgia Press, 1956.

Mount, Robert H. *The Reptiles and Amphibians of Alabama.* 1975. Reprint, Tuscaloosa: University of Alabama Press, 1996.

Perrill, Stephen A., H. Carl Gerhardt, and Richard Daniel. "Sexual Parasitism in the Green Tree Frog (*Hyla cinerea*)." *Science* 200 (1978): 1179–1180.

Phillips, Kathryn. "Where Have All the Frogs and Toads Gone?" *BioScience* 40 (1990): 422–424.

Wake, David B. "Declining Amphibian Populations." *Science* 253 (1991): 860.

Wright, Albert Hazen, and Anna Allen Wright. *Handbook of Frogs and Toads of the United States and Canada.* Ithaca, NY: Comstock Publishing, 1949.

CHAPTER 22

This project started in the same startling way as I described. (I'm truly glad that I was paying attention during Plant Pathology class thirty years before, so that I knew what was confronting me!) It is also a direct continuation of my teaching, including the "Etymological Aside" on fungal life-cycle terms.

Much good information on fungal pests can be found on the Internet, with the most up-to-date and accurate reports from state and university agricultural extension services.

CHAPTER 23

I first experienced gopher tortoises while doing consulting work for pipelines in southern Alabama. I got pretty good, in fact, at following their trails and finding their burrows. During one of these adventures, I was lucky enough to spot a huge indigo snake, one of the tortoises' burrow mates.

Good information on gopher tortoises, associated animals, and the longleaf pine ecosystem is found in scientific journals, popular articles, and government documents.

Auffenberg, Walter. "On the Courtship of *Gopherus polyphemus*." *Herpetologica* 22 (1966): 113–117.

Diemer, Joan E. "The Ecology and Management of the Gopher Tortoise in the Southeastern United States." *Herpetologica* 42 (1986): 125–133.

Iverson, John B. "The Reproductive Biology of *Gopherus polyphemus* (Chelonia: Testudinidae)." *American Midland Naturalist* 103 (1980): 353–359.

Martin, William H., Stephen G. Boyce, and Arthur C. Echternacht, eds. *Biodiversity of the Southeastern United States: Lowland Terrestrial Communities.* New York: John Wiley & Sons, 1993.

McRae, W. Alan, J. Larry Landers, and James A. Garner. "Movement Patterns and Home Range of the Gopher Tortoise." *American Midland Naturalist* 106 (1981): 165–179.

Mount, Robert H. *The Reptiles and Amphibians of Alabama.* 1975. Reprint, Tuscaloosa: University of Alabama Press, 1996.

Neal, Wendell A. *Gopher Tortoise (Gopherus polyphemus) Recovery Plan.* Jackson, MS: U.S. Fish & Wildlife Service, 1990.

Taylor, R. W., Jr. "Human Predation on the Gopher Tortoise (*Gopherus polyphemus*) in North-Central Florida. *Bulletin of the Florida State Museum* 28 (1982): 79–102.

CHAPTER 24

This chapter came from the same Arkansas field trip that engendered a previous one, "Green Tree Frogs." This time, the animal's name became a key part of the story. As with "Green Tree Frogs," my research efforts concentrated on natural history and taxonomic and ecological literature.

Altig, Ronald. "Food of *Siren intermedia nettingi* in a Spring-Fed Swamp in Southern Illinois." *American Midland Naturalist* 77 (1967): 239–241.

Barbour, Roger W. *Amphibians and Reptiles of Kentucky.* Lexington: University Press of Kentucky, 1971.

Bishop, Sherman C. *Handbook of Salamanders: The Salamanders of the United States, of Canada, and of Lower California.* Ithaca, NY: Comstock Publishing, 1943.

Caldwell, Richard D., and W. Mike Howell. "*Siren intermedia nettingi* from Alabama." *Herpetologica* 22 (1966): 310–311.

Conant, Roger. *A Field Guide to Reptiles and Amphibians of the United States and Canada East of the 100th Meridian.* Peterson Field Guide Series, 12. Boston: Houghton Mifflin, 1958.

Davis, William B., and Frank T. Knapp. "Notes on the Salamander *Siren intermedia.*" *Copeia* 1953: 119–121.

Duellman, William E., and Linda Trueb. *Biology of Amphibians.* New York: McGraw-Hill, 1986.

Dundee, Harold A., and Douglas A. Rossman. *The Amphibians and Reptiles of Louisiana.* Baton Rouge: Louisiana State University Press, 1989.

Gehlbach, Frederick R., and Braz Walker. "Acoustic Behavior of the Aquatic Salamander, *Siren intermedia.*" *BioScience* 20 (1970): 1107–1108.

Mount, Robert H. *The Reptiles and Amphibians of Alabama.* 1975. Reprint, Tuscaloosa: University of Alabama Press, 1996.

Noble, G. K. *The Biology of the Amphibia.* 1931. Reprint, New York: Dover, 1954.

Noble, G. K., and B. C. Marshall. "The Validity of *Siren intermedia* LeConte, with Observations on Its Life History." *American Museum Novitates* 532 (1932): 1–17.

Reno, Harley W., Frederick R. Gehlbach, and Robert A. Turner. "Skin and Aestivational Cocoon of the Aquatic Amphibian, *Siren intermedia* LeConte." *Copeia* 1972: 625–631.

Reno, Harley W., and H. H. Middleton III. "Lateral-Line System of *Siren intermedia* LeConte (Amphibia: Sirenidae), During Aquatic Activity and Aestivation." *Acta Zoologica* 54 (1973): 21–29.

CHAPTER 25

In this case, the puns came first. I had accumulated quite a long list of morel/moral puns when W. Mike Howell presented me with a perfect close-up portrait of the fungus, prompting me to start writing in earnest.

Because they are so widespread, mysterious, oddly formed, ephemeral, delicious (in some cases), and deadly poisonous (in others), fungi have been treated in many natural history books, field guides, and popular accounts. They gain special attention whenever they "flush," with thousands suddenly appearing.

Hudler, George W. *Magical Mushrooms, Mischievous Molds.* Princeton, NJ: Princeton University Press, 1998.

Lincoff, Gary H. *National Audubon Society Field Guide to North American Mushrooms.* New York: Alfred A. Knopf, 1981.

Lipske, Mike. 1994. "A New Gold Rush Packs the Woods in Central Oregon." *Smithsonian* 24 (January 1994): 34–45.

Paquette, Ernie. "Beauty and the Feast." *Nashville Life* (April/May 1995): 18–19.

Schaechter, Elio. *In the Company of Mushrooms: A Biologist's Tale.* Cambridge, MA: Harvard University Press, 1998.

Weber, Nancy Smith, and Alexander H. Smith. *A Field Guide to Southern Mushrooms.* Ann Arbor: University of Michigan Press, 1985.